Mankind's attempts to learn about aspects of th   its
functions by observation ar   nd
throughout history, especially   an
body was considered sacrosan  es
the origins and later develop   d
nineteenth centuries, of compa   p
with medicine and veterinary m   ...s practitioners
to understand and control outbreaks of infectious, epidemic diseases in
humans and in domestic animals. It discusses the rise and fall of the
Brown Institution and describes how it was soon overtaken in
importance by the great institutes, funded by private fortunes.

Animals and disease

An introduction to the history of comparative medicine

Animals and disease
An introduction to the history of comparative medicine

LISE WILKINSON

Royal Postgraduate Medical School, Hammersmith Hospital, London

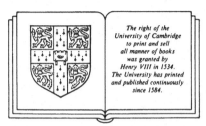

The right of the
University of Cambridge
to print and sell
all manner of books
was granted by
Henry VIII in 1534.
The University has printed
and published continuously
since 1584.

CAMBRIDGE UNIVERSITY PRESS
CAMBRIDGE
NEW YORK PORT CHESTER MELBOURNE SYDNEY

CAMBRIDGE UNIVERSITY PRESS
Cambridge, New York, Melbourne, Madrid, Cape Town, Singapore, São Paulo

Cambridge University Press
The Edinburgh Building, Cambridge CB2 2RU, UK

Published in the United States of America by Cambridge University Press, New York

www.cambridge.org
Information on this title: www.cambridge.org/9780521375733

© Cambridge University Press 1992

First published 1992
This digitally printed first paperback version 2005

A catalogue record for this publication is available from the British Library

ISBN-13 978-0-521-37573-3 hardback
ISBN-10 0-521-37573-8 hardback

ISBN-13 978-0-521-01844-9 paperback
ISBN-10 0-521-01844-7 paperback

Contents

Preface

Man's attempts to learn about aspects of the human body and its functions by observation and study of animals are to be found throughout history, especially at times and in cultures where the human body was considered sacrosanct, even after death. For this reason, comparative anatomy and comparative physiology have made relatively early appearances as established disciplines in the history of medicine. 'Comparative medicine', in the sense in which it was applied when the first university chair in the subject was established in France in 1862, came centuries later, although its roots can be perceived long before that, in the interactions of medicine and veterinary medicine in their common quest for ways of combating epidemic diseases in animals and in man.

It has been the aim in the present volume to examine the study of animal disease *per se*, for its own sake, as well as in its implications for human medicine, at first empirically and, later, by deliberate use of animal models. An added incentive has been recognition of the apparent paradox of London acquiring an institute devoted to the study of comparative medicine well before the founding of the great institutes in Germany and in particular in France, where comparative medicine first became an established, scientifically and politically recognised, branch of medical science. Hence the text attempts to examine the reasons for the surprisingly early rise, and the sadly less surprising rapid decline, of London's Brown Institution, and for the subsequent far more successful founding, and sustained growth, of other institutes with similar aims both at home and elsewhere in Europe and further afield.

I am indebted to many friends at home and abroad for encouragement and discussion, and I would like to thank the ever helpful members of staff of the libraries of the Royal College of

Veterinary Surgeons, the Royal Society, the Royal Society of Medicine, and last but not least the Wellcome Institute, who have smoothed my path in the quest for the more obscure volumes and journals and, in the case of the latter, also for suitable illustrations.

CHAPTER I

Attitudes to animal health and disease in the ancient world

The first chair of comparative medicine was established in France in 1862, a direct result of the vision of Littré and of the ambitions of Louis-Napoleon, then half-way through his ill-fated term as Napoleon III, Emperor of France[1]. But the subject represented here for the first time as an academic discipline was not new; it had been in existence for centuries, ever since infectious diseases had first begun to affect both man himself and the herds he had learned to domesticate.

Somewhere along the path of his evolution early man was forced by circumstance to begin supplementing his vegetarian diet with meat and to become a nomadic hunter. In the cave paintings discovered in south-western France and in north-eastern Spain within the past hundred years may be found the first pictorial evidence of a relationship between palaeolithic man and the large animals on which he depended for food. The relationship was between hunter and hunted. The animals depicted in the Spanish caves at Altamira are magnificent specimens in their prime, probably votive pictures painted in an attempt to persuade the powers-that-be to place such desirable victims within reach of the hunter. In the French caves in the Dordogne this line of thought was extended to include animals apparently caught in traps, or riddled with arrows[2].

Cro-Magnon man and his fellow Europeans of the Upper Palaeolithic, inhabiters of the caves and executors of the paintings between 20 000 and 15 000 BC, appear to have been physically little different from modern man, and culturally only a few steps behind. They began to bridge the remaining gap when they learned, instead of relying on nature's provisions and attempts at mystic persuasion, to provide for themselves by deliberately growing crops and domesticating potentially useful animals. With this achievement man emerged from palaeolithic wilderness into the beginnings of culture and civilisation; and having domesticated certain species of animals he was confronted with the problems of their care in sickness and in health.

It can have been no sudden development. Herds of gregarious species of sheep, cattle and horses must have built up on the steppes of Eurasia long before man formed sizeable, stable communities. Sheep and goats were the prey of packs of wolves and wild dogs, and man may have begun adding meat to his diet when he shared the kill of the canine packs. Around 12 000 BC man and dog began hunting in unison; eventually, around 9000 BC, a form of nomadic domestication of sheep and goats would seem to have been established[3].

No doubt man's adventures in therapy had humble beginnings, perhaps removing thorns from the paw of his dog, or from the hand of his woman. Problems connected with infectious disease are not likely to have existed before the inexorable advance of civilisation caused certain densities of men and animals to exist together in certain places for certain lengths of time, thus providing suitable concentrations of suitable hosts for invading micro-organisms[4]. Once such problems had become part of the daily life of an established community, medicine and veterinary medicine alike had acquired a *raison d'être*. It seems reasonable to assume that their origins are closely entwined and that when medicine first became a profession – and the definition of both the terms 'medicine' and 'profession' requires careful consideration in this context – the practitioner meted out treatment, such as it was, both to man and to his domestic animals. Probably specialisation of a kind soon ensued, perhaps even to some extent as a function of the pressure of growing numbers in both categories needing treatment. In such records as survive there are signs of differentiation in varying degrees in all the ancient civilisations. On the other hand, appeals for medicine to remain one, whether applied to man or to other animals, have been frequently heard in modern times; along the way there have been ethical and theoretical difficulties. The effects of Plato's concept of the absolute superiority of the human soul are discussed below. Later, the lack of education or even training of those who attended domesticated animals in the Europe of the Middle Ages and of early modern times led to their being viewed with disregard and mistrust. Consequently members of the medical profession sometimes found it prudent to distance themselves from any connection with animal disease in order to protect their reputations.

Domestication of wild cattle species probably began in Asia and Europe between 5000 and 3000 BC[5]. Early extant evidence of animal disease, from China and India through the Middle East to Egypt and Italy and as far north as Scandinavia, suggests that generally accepted philosophies of disease were rooted in theurgy and natural magic in all the major cultures eventually emerging during the pre-Christian era. In particular, infectious disease in epidemic form of man and of beast continued for centuries to be regarded as unavoidable punishment for past sins sent by the particular deities worshipped by the people afflicted; hence, only prayer and devotions could effect a cure. As late as 1871, five years after the last disastrous importation of cattle plague into the British Isles, George Fleming could write:

> On the Continent, St. Cornelius and other saints of France, and St. Antonio of Rome and Italy in general, are the protectors of four-footed creatures; and it is much less troublesome for their owners, and more profitable to the priests, to obtain exemption from an approaching plague through the merits of a mouldy saint than by the adoption of onerous, heretical measures of a hygienic kind, which do not benefit the Church[6].

But alongside the incantations and other stock-in-trade of those attempting to ward off epidemics and epizootics there emerged in the course of time, by the application of a mixture of common sense and empiricism, a realisation of the necessity for measures of isolation in cases of epizootics among domesticated animals, and to combat certain contagions in man. Even if the initial outbreak were still considered divine punishment, man could attempt to limit the extent of damage when contagiousness was an obvious possibility. Documentary evidence for this type of development is found in many early cultures[7].

According to Herodotus, who wrote in the fifth century BC,

specialisation in Pharaonic Egypt had evolved to a point where different physicians treated complaints of different organs. Some nineteenth century historians claimed that veterinary practitioners, although belonging to the lowly shepherd class, were highly skilled in the art of treating animal disease, and that each practitioner treated one animal species only. The latter claim has later been shown to be open to question and may be due to a mis-interpretation of documentary evidence[8]. The few fragments left of the Kahun Papyrus seem to suggest that the animal healers, no less than the physicians, of ancient Egypt may have depended largely on the supernatural rather than on any attempts at rational therapy. Paradoxically, because of resistance to post-mortem examination, the practice of embalming man and animals did nothing to further the acquisition and recording of knowledge of anatomy and pathology. Of the three surviving, incomplete, treatments recommended in the Kahun Papyrus, one appears to have been aimed at curing a condition akin to influenza in cattle; it consisted chiefly of incantations, sprinklings with cold water, and rubbing with gourds or melons. On the other hand, Schwabe has recently pointed to the effects of observations of muscle contractions during bull sacrifice in pre-dynastic Egypt, believing such observations to represent the beginnings of an evolution of scientific methods from the practices of healing magic[9].

With regard to the evolution of a separate art of veterinary medicine, India among the ancient civilisations probably made the most distinguished contribution. The comparatively high standard of early Indian veterinary practice is hardly surprising in view of the nature of Hindu thought and Hindu religion; the belief in reincarnation imbued every Hindu with a veneration for all animal life. Even before this belief was adopted (between 800 and 200 BC) the necessity of caring for such animals as were important in the domestic economy had produced the beginnings of veterinary science. Kings and princes as well as physicians were concerned with problems of animal health, and extant medical treatises contain chapters on the care of healthy as well as of diseased animals. Two veterinary counterparts to the *Charaka samhita* of more than 10000 verses each are the *Haya-Ayurveda*, on horses, and the *Hasti-Ayurveda*, on elephants. At the time they were written, around 800 BC, physicians treating human beings were also trained in the care of animals, although some practitioners specialised in animal care exclusively, or even in the care of one class of animals only. Hospitals for animals were provided alongside the State hospitals for needy human beings as early as the third century BC[10].

In China, early dynasties possessed animal healers and had specific titles for healers of man on the one hand, and healers of all other animals on the other, as had also the Sumerians and the Babylonians, who eventually reached a level of specialisation similar to that found in Egypt. In China as in most other civilisations the horse had a major share of veterinary attention. A manual of equine diseases was written before 2000 BC, as was a counterpart devoted to diseases of the domestic buffalo, perhaps reflecting a more equal and lasting importance here of the two species side by side than further west, where the bull–cow culture of Egypt lasted unchallenged for four millennia before being replaced by societies focusing on the horse. In the second century AD, one Roa-Tro perfected an isotherapy of his own invention involving treatment with diluted sweat of domestic animals susceptible to the same disease as the afflicted human. Through the centuries Chinese disease concepts have differed radically from western ideas[11].

As in other areas of our history it is to the Graeco-Roman culture that we must look for the more immediate roots of our own attitudes with regard to medicine and veterinary medicine. The Greek legends had Asklepios heal both animals and men. Leclainche[12] has pointed out the effects of the Platonian philosophical tenets on contemporary Greek medicine, and on Plato's near contemporary Hippocrates and the Hippocratic writings. Plato's concept of the immortality of the human soul, and its kinship to divinity, was contrasted by a low opinion of the inhabitants of the animal world. He could not approve of comparisons between human and animal life, or even of observation of animal behaviour.

The writer or writers of the Hippocratic corpus did not wholly accept the Platonian *tabu* on animal observation, and although generally contemptuous of veterinary medicine found comparative pathology useful. At the time of the Italian outbreak of cattle plague in 1711–13 (see chapter 3) both Ramazzini and Lancisi referred to the fact that the great Hippocrates had not found observation of cattle beneath his dignity. Lancisi included more than a page of references and quotes from the Hippocratic corpus to illustrate the comparisons made with plant and animal worlds, concluding with Hippocrates' reference to bursting vesicles as the cause of chest pain in dropsy: 'I have evidence in the ox, for if you cut him up, you will get a sure understanding'. Hence dissection was not avoided, and sometimes encouraged, when comparisons were considered useful and of constructive value[13].

Exactly what and how much of the Hippocratic corpus was written by Hippocrates of Cos remains open to question, although the first and third book of *Epidemics* are among the favoured attributions. These books relate prevailing climatic factors to observed outbreaks of fevers, consumption, etc., and describe individual cases, often with a fatal outcome, of named patients. Many of the diseases described in endemic and epidemic form appear to have been closely related to infections abroad today. *Airs, Waters, Places*, on the other hand, discusses in more general terms the effects of differences in the seasons, and of different climatic and geographical factors, on the health and development of populations in Europe and in Asia. Meant as guidelines for the itinerant physician, who might need to adjust to changing situations, the text of *Airs, Waters, Places* addresses problems of epidemiology in a wider context. The writer considers the development and state of health of domestic animal herds and the crops on which they all depend, as well as that of human populations[14]. Some writers on veterinary history have added to the confusion over authorship of the Hippocratic corpus by attributing to Hippocrates certain treatises on animal disease; these were in fact written at a much later date, probably in the fourth century AD by another Hippocrates altogether, who in turn was not identical with one 'Ippocras' whose work, in Sanskrit, appeared in the sixth century[15].

Already with the advent of Plato's pupil Aristotle (384–322 BC) came another radical change. Initially a faithful follower of the Platonian meditative approach ready to accept a common inherent divinity of human soul and universe, Aristotle had an independence of mind which enabled him to strike out on his own and throw off the shackles of Plato's and of Socrates' inward-looking contemplation of abstract concepts. Plato had regarded the physical examination of objects as a superfluous exercise, believing that their essential mystery could be explored only through meditation. Aristotle rose above his heritage from the philosophers of his day and before, and laid the foundations of experimental natural science and even of a comparative approach. He described a number of animal species and some of their diseases; he observed the development of the chicken inside the egg. His comments were not always particularly lucid, nor were they always accurate judged by later standards; but such considerations should in no way be allowed to detract from the reputation of this highly original trail-blazer in experimental science[16].

Aristotle's thoughts on infectious disease were few and splendidly idiosyncratic. He suspected the shrewmouse and certain small lizards of injuring pastured horses by venomous bites. As a result, the shrewmouse was accused for centuries afterwards of being responsible for sudden deaths of animals at pasture, which may in many cases have been caused by anthrax[17]. He also described diseases of cattle which were almost certainly foot-and-mouth disease and pleuropneumonia, without touching on questions of control. He referred to the lungs of affected animals as being 'spoiled', which must imply the practice of post-mortem examination. His brief remarks on rabies have been often quoted, largely because of the puzzling aspects of his statement that 'Rabies drives the animal mad, and any animal whatever, excepting man, will take the disease if bitten by a mad dog so afflicted; the disease is fatal to the dog itself and to any animal it may bite, man excepted'[18].

Aristotle was far from being a practising physician or veterinarian. His thoughts on infectious disease offer little illumination, but his animal studies were seminal for the comparative studies in anatomy and physiology initiated during the third century BC in the two rival schools founded in Alexandria by Herophilus and by Erasistratus. We have no first-hand knowledge of these two authors, whose comparative studies appear to have involved, more often than not, criminals as well as animals; their work perished in the burning of the great library at Alexandria in AD 391[19]. Among the Romans in the pre-Christian era the celebrated statesman Cato the Elder (234–149 BC) did not improve his image as moral reformer by writing a treatise on agriculture. He put profit first, and his only advice to owners of diseased animals and slaves was to get rid of them to the highest bidder, and failing that to drive them off into the wilderness, to prevent them being a burden.

We owe most of our information concerning this period to the accounts left by Celsus and by Galen. Celsus' works also are incomplete; although we have the eight books of *De medicina* to enable us to form an opinion of the state of Roman medicine in this age, their counterparts concerned with domestic animals have not survived. There were five books on agriculture mentioned by later authors. It is in *De medicina* that Celsus refers repeatedly to the works of Herophilus and of Erasistratus and incidentally tells us that according to his sources the two sages studied the human body 'in the best way so far' when they 'laid open men whilst still alive – criminals received out of prison from the kings'. His own advice on wounds and bites was sensible and, with hindsight, fortuitously apt. He recommended cupping to draw out the 'virus' from dog bites 'especially if the dog was mad'; also cauterising the wound if at all possible; and described the distressing illness and death of patients when treatment had been ineffective[20].

Celsus, in the first century, also made an occasional critical remark concerning the work of Erasistratus and Herophilus; Galen (*c*. AD 130–200) was merciless in his criticism of Erasistratus in particular. His own work on comparative anatomy and experimental physiology relied on dissection, and vivisection, of large numbers of different species of animals; by Galen's time, dissection of the human body was no longer acceptable. Singer called Galen's anatomy 'that of the soft parts of the Barbary ape, *Macaca inuus*, imposed on the human skeleton'. Among other species used experimentally by Galen were pigs, sheep, cattle, cats and dogs; and on occasion weasels, bears, mice and even an elephant[21].

Probably, as was the case in earlier and in later works, Celsus' lost books on agriculture and husbandry placed more emphasis on the care and treatment of the horse than of other species, although Praximus, who wrote around 100 BC, is said to have described cattle diseases. Among the works lost in the obliteration of Carthage were Mago's 28 books on agriculture and a volume on

Fig. 1. Galen demonstrating vivisection of pig, from the 1556 Venice edition of the *Opera omnia* (courtesy Wellcome Institute Library, London).

diseases of dogs written by one Aurelius Olympius Nemesianus early in the third century BC[22].

However, from the time of the Romans and until the establishment of veterinary schools in Europe coincided with devastating outbreaks of cattle plagues in the eighteenth century, writers on veterinary matters paid scant attention to animal species other than the horse, equally important in war and in peace and not least as a status symbol to the aristocracy. Most of these writers belonged to the landed gentry, many of them accomplished horsemen themselves with a proprietary interest in husbandry. Farriers and cow-leeches were for the most part illiterate, although the views held, and sometimes the therapy used, by certain 'Court Veterinarians' occasionally found their way into some of the works.

The presence of a number of infectious diseases of domestic animals is documented in descriptions throughout Graeco-Roman literature (what is preserved of it) from Aristotle to Pliny and, later, Vegetius (*fl.c.* AD 450). A clear picture rarely emerges; among the symptoms mentioned some suggest, in the horse, glanders, farcy, and strangles, and in cattle, contagious bovine pleuropneumonia and rinderpest. Treatment, where any was recommended, may today appear surprising, even bizarre. In their own time, remedies such as concoctions of blood of sea-tortoises and spiced wine poured through the nostrils of ailing cattle, or the elaborate manner of applying root of hellebore, or of lungwort, to incisions in the ears of sick animals, probably held out as much hope of a cure as many less exotic ones[23].

The more enlightened writers, among them the erudite polymath Varro (first century BC) gave a measure of common-sense advice. Varro suggested keeping animals in small separate groups as a means of avoiding large-scale destruction in the event of major plagues. He seems to have envisaged isolation of animals as a preventive measure at a time when writers of the Bible were wrestling with the problem of reconciling the desire to isolate human sufferers from 'leprosy' with the dictates of Christian compassion and attempts to heal[24].

A great deal of attention has been lavished on Varro's introduction into the literature of a concept of a world of tiny invisible animals 'carried with the air into the body by way of the mouth and nostrils, giving rise to serious diseases'[25]. Varro's views could well represent a compilation of the theories of other writers before him, since his works on agriculture have survived where many others, contemporary and earlier alike, have been lost. It will never be known for certain who first introduced the concept of tiny invisible beings as carriers of disease into the literature. What is certain is that Varro, author of *De re rustica*, was no more a veterinarian or agriculturalist than Celsus, author of *De medicina*, was a physician. A later age would have labelled them natural philosophers, or encyclopaedists.

The encyclopaedic tradition continued throughout the Roman period. The works of Virgil (71–19 BC) and of the Elder Pliny (AD 24–79) bear witness to the familiarity with major plagues of animals and of man which was forced on the inhabitants of the Roman territories, and perhaps in the end contributed to the decline of the empire. Pliny's contemporary, Columella (*fl.* AD 50) completed a treatise on husbandry about AD 55, possibly based to some extent on Celsus' lost works on agriculture. He wrote more fully on animal plagues than other known writers of the period, and impressed on his readers the importance of segregating the sick from healthy animals to prevent the contagion spreading; there is no mention of destroying those affected.

Vegetius, who wrote in the fifth century AD, was the last of the Roman encyclopaedists. His treatise on veterinary medicine was published both in Latin and in German before the middle of the sixteenth century. This early appearance, in terms of printed works, has occasionally earned for Vegetius the title of father of veterinary medicine, an appellation hardly justified in view of the highly derivative nature of his writings. Even the identity of Vegetius has long been open to question; for centuries opinions have varied as to whether or not the Vegetius Renatus who wrote *Ars veterinariae* was identical with the Vegetius Renatus who compiled the *Epitoma rei militaris* dedicated to 'The Emperor', probably the great Theodosius I of Byzantium. In a foreword to the first Latin edition printed in Basle in 1528 the publisher made it clear that he considered the author of the veterinary and of the military treatise to be one and the same man. After four centuries of changing persuasions, current opinion favours the same view[26]

Later critics have differed in their judgement of the value of

Fig. 2. Woodcut from the title page of Joannes Ruellius's Latin version (1530) of the *Hippiatrika* (courtesy Wellcome Institute Library, London).

Vegetius' contribution. Leclainche dismissed him as 'not erudite' and having but a poor comprehension of the classical works of Hippocrates and of Galen, and of the Greek philosophers, although he gave him credit for compiling the definitive account of the opinions of his era. Several other commentators were even more critical; but Sir Frederick Smith, after a lengthy analysis, concluded that while Blaine had exaggerated in proclaiming Vegetius the 'veterinary Hippocrates', his remarkable work had been insufficiently recognised by the veterinary profession. Chiodi, the most recent author to examine Vegetius' veterinary work at some length, objectively considered the good and bad points of the volume. Placing Vegetius in his historical context in an age of decline, when medicine itself suffered and was to continue to suffer for several more centuries from the absence of worthy successors to Aristotle, Hippocrates and Galen, Chiodi considered him to have made a laudable effort to integrate current medical thought in an honest attempt to create sensible guidelines for the care of domestic animals[27].

Whatever his identity, and derivative (and hence representative of more general views) though his writing may be, Vegetius did not confine himself to the field of hippiatric medicine. What he did write on the subject owed much to the contributions by Apsyrtus and by Pelagonius to the *Hippiatrika*, that monument to Byzantine care for the horse, first compiled a century earlier[28]. His third book of four is devoted to the care of cattle, and in a chapter headed '*De morbis boum et primo de malleo*', which may have been suggested by the reading of Chiron rather than of Apsyrtus, Vegetius left us a description of cattle plague, or rather of cattle plagues. He used the generic term '*malleus*' for epizootics of different kinds of both horses and cattle. In the case of the latter, Smith discussed the identification of a '*malleus*' in which there was a profuse flow of saliva, unable to decide between foot-and-mouth disease and rinderpest. Fleming in 1871 pointed out that 'The nature of this malady cannot be accurately enumerated, but it is obvious that several affections are included in this general designation'[29].

Columella had been the first to state that 'The diseased must be separated from the sound, that not so much as one may come among them which may with the contagion affect the rest'. Vegetius wrote more loquatiously although with impressive clarity, and allowing himself a swipe at dangerous superstition:

> All these diseases are very contagious, and if one animal be seized by them they pass immediately to all; and so they bring destruction either upon whole herds or upon all those that are fully domesticated and trained to labour. Therefore it is that the animals which have been attacked must, with all diligence and care, be separated from

Flauij Vegetij Renati Ain

Büchlein / vonn rechter vnnd warhaffter

kunft der Artzney/Allerlay kranckheyten / ynnwendigen vnd außwen-
digen aller Thyer/So etwas zyehen oder Tragen mügen /Als pferd/Esel /Maul-
thyer/Ochsen/vnd anderer/Auch wie man allerlay kranckhayten art vñ gepresten
erkesten soll/die mit getrencken/Salbungen/ pressungen/ Lässen/ vnd ander Artz
neyen etc. zůuertreyben/Vormals durch Vegetiū Renatū in Latein beschriben/
yetzunder/ inn Teütsche sprach verwendt/Allen vich ärtzten/Marstallern/
Schmiden/Reyttern/Burgern vnnd pawren/Auch allen denen die
mit gemeltem vich vmbgeend/gantz nutzlich vnd not-
wendig zů geprauchen.

M. D. XXXII.

Fig. 3. Title page of the first German edition of Vegetius'
Veterinary Art (Wellcome Institute Library, London; by courtesy
of the Wellcome Trustees).

the herd, put apart by themselves, and sent to those places where no animal is pastured, lest by their contagion they endanger all the rest; and the negligence of the owner be imputed (as is usually done by fools) to the divine displeasure.[30]

Vegetius recorded his views on transmission and isolation during the fifth century AD, a time of turmoil in the Western world. Greek and Roman ideas, which had held sway for so long, were being replaced by social and economic forces used by an increasingly powerful Christian Church to mould the barbaric world of the Dark Ages into a 'closed Christian society'. It was a long and painful, if ultimately successful, transformation of a semi-pagan world into mediaeval civilisation[31]. In the interim, the views of the 'fools' referred to by Vegetius were widely disseminated as a general belief in divine displeasure and preternatural phenomena as causes of epidemics and epizootics; in its established form it survived for centuries, even among the better educated sections of society, as subsequent literature testifies.

CHAPTER 2

From the Dark Ages to the dawn of enlightenment

The tentative ideas concerning aetiology and control of epidemics and epizootics found in the writings of the Greek and Roman encyclopaedists were eclipsed throughout the period of the Dark Ages. Towards the end of this period they reappeared in Arabic translations of Greek texts. In addition to the translation and preservation of surviving older texts, an important Arabic contribution at this time to the literature on infectious diseases was Rhazes' treatise on smallpox and measles. It is the first properly identifiable account of smallpox in the surviving literature, and Rhazes himself began by discussing the evergreen question of whether or not Galen had mentioned, or indeed known, smallpox. Rhazes thought he had, but was unable to explain why then such a meticulous clinical observer had nowhere given a clinical description of the disease. Rhazes' own account and his discussion of 'the causes of smallpox; how it comes to pass that hardly anyone escapes the disease' is detailed and accurate. The apparent inevitability of an attack suggests that smallpox must have been endemic in the Persia of the time, and that consequently nearly everybody suffered an attack in childhood, with subsequent immunity. This distribution, and presumably a relative mildness of the disease at the time, led Rhazes to believe that the onset in young children coincided with a stage in their development when their blood changed character, and that it was a natural and inevitable phase[1].

This belief that smallpox and measles, which he appears to have regarded as a variant of smallpox, were not only inevitable but formed part of a desirable, constitutional and very natural phenomenon, prevented him from paying much attention to possible preventive public health measures, or indeed from considering the disease in the light of major threatening epidemics; much less from comparing it with the cattle epizootics which in the time of the Greeks and the Romans had already given rise to measures of isolation and quarantine. His treatise forms the first independent Arabic contribution to the literature on infectious diseases. A century later Avicenna, in the first book of his Canon, attempted to distinguish between different paths of transmission of infections in man, and discussed the changes in the atmosphere which make the air noxious[2]. As in many another culture before and since, the Muslim healers were hampered by a ban on dissection of man. Having no Platonian reservations they turned to animal dissection which encouraged forays into comparative anatomy but did not open up the field of epidemiology.

By the thirteenth century Europe had begun the slow recovery from the Dark Ages. Some credit can be ascribed to elements within the 'closed Christian society' and paradoxically within the most closed communities of them all, the great monastic orders. The avowed vocation of the leaders of orders such as the Dominicans and the Franciscans was not just preaching and missionary work; study formed an integral part of the lives of their members. Initially their studies were not supposed to include medicine as such. Albertus of Cologne (1206?–80) joined the Dominicans shortly after the inception of the order in 1216. Before this, his early studies at Padua are thought to have included some medicine; once a Dominican, he turned to animals and their physiology and diseases. His *De animalibus*, written in the mid-thirteenth century, is a testament to his interest in natural science and Aristotelian thought. In one chapter he discussed human and animal plagues together; in another, he left us the earliest known example of a clear statement of possible pathways of transmission of animal infections. He distinguished between (1) inoculation by bites or other injuries; (2) actual contact with diseased animals; (3) respired air from the sick. Albertus himself considered the third possibility the most important one[3].

There is no reason to doubt the generally accepted assessment of Albertus as the quintessential compiler. Whatever his early leanings, medical theory formed no part of his later works. The

Fig. 4. Albertus Magnus and students as depicted in
Harderwyck's *Epitomata*, 1496 (courtesy Wellcome Institute
Library, London).

nature of his creed excluded any possibility of this, and it must be assumed that his clear-cut thoughts on the nature of spread of contagious disease represented accepted views in informed circles in the Europe of the thirteenth century. It would be difficult to claim originality for Albertus' three categories of spread of infections among herds of domestic animals; yet there are no obvious sources in any surviving manuscripts, nor in the Arabic literature preceding him. Albertus did not refer to specific diseases among cattle, and may not have written on the basis of personal observation. He is most likely to have drawn on manuscripts such as Vegetius' and the more diffuse discussion of transmission of human diseases presented in the first book of Avicenna's Canon[4]. The succinct formulation is his own.

At the time Albertus was writing and teaching at Dominican centres at Paris and Cologne, southern Italy was emerging as a centre of the new waves of learning which were to culminate in the Renaissance. Frederick II had been king of Sicily (including southern Italy) since 1197 when he was three years old. He was to prove an enlightened ruler, eventually as the Emperor Frederick II from 1220 until his death in 1250. Keenly interested in the progress of learning, his support of the existing medical school at Salerno gave it a new lease of life. It had been founded in the ninth century but had survived only as an isolated phenomenon, perhaps too far ahead of its time to develop to any great extent in the meantime. Frederick took a personal interest in all the natural sciences, and like many a contemporary nobleman he developed a special fund of knowledge of horses and hawks, writing himself a treatise on birds of prey and hawking, *De arte venandi cum avibus*.

To regenerate the art of equine surgery and medicine, Frederick appointed one Jordanus Ruffus, who appears to have been friend and adviser to the emperor on other matters as well. In producing his treatise on equine medicine, Ruffus relied heavily on the Arabic reservoirs of originally Greek knowledge, supplementing these sources with facts absorbed by study of Roman authors and occasionally by original observations. He was constantly encouraged by his royal employer, but their joint interest in veterinary science appears not to have extended beyond domestic concerns with horses and the birds of prey used for hawking and falconry. A recent biography chronicles Frederick II's interest in the writings on falconry in a chapter on 'culture at court', but does not mention Ruffus. Nineteenth century authors who analysed both the language and the contents of Ruffus' manuscript in detail pointed out that he appears not to have made use of the Greek *Hippiatrika* since the Greek knowledge of diseases, especially contagious ones, was more extensive than his[5].

Whatever Ruffus' strengths – and he has been much praised by veterinary historians – they did not lie in theoretical considerations of infectious diseases, and much less in comparative studies. A child of his time, he favoured humoral pathology and his notions of contagion were vague. He knew and competently described strangles but did not consider it a contagious disease; glanders, on the other hand, he knew to be contagious. As for isolation and destruction of infected animals, he had learned nothing from the Roman writers and recommended that infected animals be turned out to grass with their fellows to get rid of the discharge.

Other thirteenth century authors writing on animal disease include Giacomo Doria, the only copy of whose manuscript is in the *Marciana* in Venice, a distinction probably owing more to the noble birth of Doria than to the excellence of his treatise. The nineteenth century critics referred to above, essentially 'whiggish' in their historical approach, dismissed it as a 'very superstitious work', citing as an example Doria's recommended treatment for farcy, which consisted in the celebration of Mass and the consecration of three Hallelujahs. Theodore of Lucca, or Theodoric Borgognoni, a contemporary of Albertus, was born in Bologna and became Bishop of Cervia. He left a number of manuscripts during a versatile career which reflected his catholic interests and even included a suggestion of an approach to comparative thinking. The son (or disciple) of Ugo Borgognoni, whose objections to the laudable pus theory he shared, Theodore studied medicine before entering the Church, then returned to medicine, and eventually added the practice of veterinary medicine to his activities. Copies of several of his manuscripts are preserved in major Italian libraries; one concerns itself with medicine and surgery of man and of other species. Such versatility was not unusual among his Italian contemporaries[6].

Fifty years after Ruffus, Laurentius Rusius wrote along the same lines; also essentially a compiler he was a practising veterinarian who could supplement his borrowings from Ruffus and others with observations from his own experience. Later commentators have praised him for his competent rendering of the knowledge of his times, and in particular for his recognition of the possibility of wounds leading to general sepsis, and for his warning of the transmissibility of farcy[7].

The resurgence of veterinary literature in Italy during the thirteenth century can be seen as a corollary of the crusades. In the twelfth and thirteenth centuries crusaders from all over Europe converged on the roads through Italy and to Constantinople to embark for Palestine. The horses they brought with them stimulated interest in equine matters, as did the impressions of Arab horsemanship brought back by returning survivors. As always, movements of troops and their animals also brought in their wake outbreaks of epizootics; anthrax and cattle plagues appear to have been major scourges in the accompanying cattle trains at this time, and glanders was a ubiquitous presence, in and out of the armies. However, even when the industrious scribes of the Middle Ages wrote on disease problems of several species, there was little attempt to benefit from comparisons in the field of infectious disease, although comparative anatomy and to some extent physiology had been practised with some success from the time of the Greeks. Understanding of infectious disease processes, in all and any species, was and largely remained rudimentary until the nineteenth century.

Whether or not they were inter-related, the ideas of Avicenna and of Albertus stand out as the clearest formulations of concepts of contagion and transmissibility before the fourteenth century. But the dawn of scholarship in Europe represented by the activities of Albertus and his contemporaries in the Dominican and other orders was interrupted soon after Albertus' death in 1280. From the early years of the fourteenth century the incipient cultural renaissance heralded by Albertus and others suffered an eclipse. The fragile economy of Europe was shattered by a series of natural catastrophes. Volcanic eruptions in Italy and devastating floods along the course of the Rhine were followed by crop failures in France. Then outbreaks of cattle plagues throughout Europe, not excluding the British Isles, resulted in widespread famine. By mid-century the populations already weakened in such ways suffered the worst blow of all. In the spring of 1347, bubonic plague was introduced to Constantinople from Asia. To judge from not always consistent accounts it would seem that waves of bubonic and pneumonic plague were attacking hosts whose resistance was low, and between 1348 and 1350 the Black Death devastated the European continent. There is no doubt that other infectious diseases also played a part. It has been pointed out that typhus, influenza and smallpox may all have contributed to this greatest single demographic disaster, while anthrax, rinderpest and other cattle plagues attacked domestic animals; but the progress of the Black Death among weakened populations with low resistance and little immunity obscured other misfortunes as it decimated communities and disrupted social structure. From this catalogue of unprecedented disaster emerged the early plague tractates written between 1348 and 1350 by a number of authors, some of whom later succumbed to the disease. Winslow examined them all in some detail and concluded that the writers were united in their conviction that the plague was highly contagious (this would fit pneumonic plague better than the bubonic form) and that the only hope of avoiding it was to avoid infected areas. They were also united in their belief that the contagious principle transmitted an only vaguely identified, mysterious property present in the atmosphere when it was corrupted by a variety of influences, from exhalations from bodies of the dead and other 'putrefactions' and vapours formed in the interior of the earth to climatic factors and, above all, unfortunate constellations of the stars and planets[8].

One author who lived through the Black Death and tended its victims was Guy de Chauliac, physician to three successive Avignon popes, from Clement VI in 1348, the year of the onset of the Black Death in Europe, to Urban V in 1363, the year he wrote his *Grande Chirurgie*. By the time he published his reflections on the great epidemic he was in a commanding position to do so; he had observed its progress, treated its victims and, at the end of the outbreak, he himself had been fortunate to survive a serious attack of the disease. His observations include clear distinctions drawn between the lesions of bubonic plague and those of other diseases accompanied by similar lesions, including anthrax. It is a first-hand account, which provides a refutation of more recent attempts to suggest alternative identities for the pestilence known as the Black Death[9].

There was little new literature on plagues of animals in this century so preoccupied with the more immediate disasters of epidemics of humankind; yet the speed with which the earliest plague tractates appeared and gave common-sense advice on quarantine and isolation measures may suggest that the lessons learned from epizootics in earlier centuries were quickly reapplied to problems of human disease. Only the most effective measure, the ultimate solution, was unthinkable in this new context. However valuable, animals can be slaughtered in efforts to contain the spread of infection; infected human beings may be transferred to isolation wards and hospitals but must always be nursed by fellow men and women who may themselves be susceptible to the infection.

For almost two centuries after the Black Death there was scant progress in epidemiological literature and in public health management. Manuscripts concerned with human and animal disease were mostly compilations of the works of previous centuries with little originality and often with gratuitous introduction of fresh superstition. Towards the end of the fifteenth century the situation began to improve. The art of printing had been invented. As a result new literature as well as important manuscripts of previous centuries were being printed and made available for wider distribution. Throughout Europe science, including medical and veterinary science, began to develop and emulate the renaissance in art and architecture which had been evolving steadily from its centre in Italy since the latter half of the fourteenth century. In an effort to inform a vulnerable public, concerned authorities also began to issue broadsheets designed as guides to common diseases of horses and cattle for use in farming communities (Fig. 5). Examination of such broadsheets reveals the presence of a number of diseases still causing concern centuries later.

The most significant advance within the field of medicine during the first half of the sixteenth century was a very basic one, essential for further general progress. It came with the publication, in 1543, of Vesalius' *De fabrica humani corporis*, and with it reality entered the study of human anatomy even for those who did not have the good fortune to observe the dissection of human cadavers. Such dissections had been performed more or less clandestinely at Padua and Bologna, although at Montpellier the activity had been taking

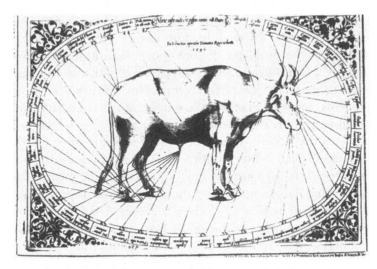

Fig. 5. Engraving of ox surrounded by 45 numbered 'seats' identifying 45 different diseases. 'Fugitive sheet' of veterinary information issued at Venice in 1596 (courtesy Wellcome Institute Library, London).

place sufficiently publicly to be illustrated in certain fourteenth century manuscripts such as that of Guy de Chauliac. Elsewhere, students had relied largely on veterinary observation, a practice which could, and sometimes did, result in erroneous conclusions drawn from a one-sided exercise in comparative anatomy. In spite of difficulties presented by the hostility of religious and political adversaries, Vesalius and other dedicated observers at this time set a standard for the study of human anatomy which has stood the test of time.

Three years after the appearance of this epoch-making work Girolamo Fracastoro published in Venice *De contagione et contagiosis morbis et eorum curatione*. Vesalius' work was the sublimation of a number of attempts to come to terms with the anatomy of animals and man; by virtue of its title and its concept of 'seeds of disease' Fracastoro's has often been regarded as an altogether new departure. The validity of such claims has recently been re-examined by Nutton, who considered possible sources for Fracastoro's concepts. He concluded that 'Fracastoro's claim to greatness cannot lie in the originality of his conception of seeds of disease or of contagion', but that his lucid and elegant prose had made him an ideal expositor of ideas from antiquity to his own day[10]. Certainly his was the first known treatise to be devoted entirely to the problems of contagious disease; it was also a first attempt to consider together infections of animals and man, and even to draw on plants for comparative purposes in its general chapters. Some of Fracastoro's writing may have been inspired by extensive outbreaks of foot-and-mouth disease, which began in Friuli in the Veneto, spread in a south-westerly direction through the Eugenaean Hills south west of Padua, and eventually threatened the area around Lake Garda where he owned property. Describing the symptoms and the course of the disease, Fracastoro added a warning that '...the infected animal had to be removed at once from the herd, or the whole herd became infected', which is not necessarily more than an echo of Vegetius. But Fracastoro was interested in the whole spectrum of infections, with particular emphasis on contagious diseases in man. His reference to comparable transmissions of disease in plants and animals seems to be wholly his own, although he did not elaborate on the comparative aspects and merely gave his examples of analogy in a matter-of-fact way. He wrote of pests spreading from grape to grape, and from apple to apple, in the manner of all contagion, choosing from his examples crops of economic importance. His most explicit comparisons were reserved for his comments on specificity when he wrote of 'a certain pest which attacks trees and crops, but harms no sort of animal; again, there is a pest which attacks certain animals but spares trees and crops. In the animal world, one pest will attack man, another cattle, another horses, and so on ...'[11].

The paragraph on the cattle disease in the district of Friuli followed a discussion of differences between scabies, phthisis, and 'pestiferous and lenticular fevers' after Fracastoro had in previous chapters distinguished between contagion and putrefaction, and also between poisons and contagion. His closing remark on what appears to have been an outbreak of foot-and-mouth disease reads: 'Gradually this distemper descended to the shoulders and thence to the feet, and in cases where this change took place almost all the animals were cured but when it did not, they nearly always died'. This sentence could just suggest that rinderpest, with its much higher case fatality rate and no lesions on the feet, might have been present as well. That was the extent of Fracastoro's use of animal disease for comparison. He went on to discuss the origin of such contagions, concluding in time-honoured fashion that 'the air is the most potent cause, though they may also come from water, marshes and other sources'. As for his ideas on the spread of contagion by means of 'seeds' (*seminaria*) of disease, Winslow remarked many years ago that he 'worked out a clear and essentially accurate analysis of the way in which living "germs" operate, without ever suspecting that they were living'[12].

Suspicion that disease agents might be living organisms began to emerge in the following century, when bubonic plague continued its periodic, remorseless run as a great protagonist of tracts on contagion, its nature and the possible means to combat it. Athanasius Kircher (1602–80) wrote the seventeenth century companion piece to Fracastoro's treatise on contagion; but Kircher was concerned with bubonic plague only, and he lacked to a singular degree Fracastoro's gift for clear and instructive writing and consideration of the problems at hand. Although it is thought unlikely that Kircher could have seen actual micro-organisms, he probably saw red blood corpuscles in the tissues he examined; whatever he saw, he was able to make the leap from 'atoms to organisms' and interpret the structures as living, as an *effluvium animatum* of plague contagion. There had been precursors to this view earlier in the century, and indeed in the previous century; but Kircher's *Scrutinium Pestis* of 1658 was the first work to argue in depth for the existence of living agents of infectious diseases on the basis of microscopic observation[13], however questionable its accuracy. He did not argue very clearly, and his arguments were probably based on a misinterpretation of what he had seen under a relatively primitive microscope. Nevertheless, his *Scrutinium Pestis* stands as the first serious formulation of the case for the independent life of agents of disease. He believed that putrid effluvia released from below ground rose to infect plants, animals and man, and that once having invaded these hosts they were able there to give rise to *effluvia animata* which in the form of tiny worms caused disease. The 'worms' he thought he had seen in the blood of plague victims seemed to support his theory. He also graphically described an epizooty at Venice earlier in the century which Fleming in 1871 redefined as 'angina maligna, a form of anthrax...even transmissible to man'[14]. Kircher himself related cases of herdsmen and farmers eating the flesh of infected animals and 'succumbing to the noxious food'.

Kircher's reference to the Venice outbreak reflected a growing concern with aspects of agriculture in northern Italy at this time. A near-contemporary of Kircher was Agostino Gallo (1499–1570), a native of Brescia, who did much to promote the cultivation of rice in the Po Valley, and who also included diseases of domestic animals among his agricultural concerns. It is clear from his writings that contagious bovine pleuropneumonia was a considerable problem for livestock farming in mid-sixteenth century northern Italy; together with rinderpest and to a lesser extent foot-and-mouth disease its presence was to remain a continual threat, growing to disastrous proportions over the next century and a half. Gallo had no faith in remedies; only immediate and complete isolation of infected animals could halt the spread of this 'desperate' disease to healthy stock. He also advised washing the cattle's feed with water boiled with scented herbs[15].

The *Scrutinium Pestis* was first published at Rome in 1658, approximately halfway through a century which politically and historically was fraught with upheavals and strife in several European countries. The Thirty Years' War and its ramifications, that gigantic struggle for power between the Austrian Hapsburgs and the German princes, coincided with a general change in the intellectual climate of Europe. The Renaissance was making its final progression from the arts to the sciences, and from Italy into the rest of Europe. At the same time, the war and the phenomena and crises it brought in its wake, its political and religious persecutions, caused an intellectual migration from the European continent comparable to the Huguenot exodus from France following the 1685 revoking of the Edict of Nantes and also to that caused 300 years later by the Hitler regime and the ensuing World War II. In all cases it was the Anglo-Saxon world which was the main beneficiary of the influx. During the first half of the seventeenth century a number of Palatinate expatriates came to England, and many of them settled in London and had become a considerable intellectual influence by mid-century. By then perhaps they had begun to question the wisdom of their choice. The Civil

War, the execution of Charles I, the events of the Protectorate hardly improved the lot of the scholar in their adopted country. On the other hand, during this period London became the venue for a great deal of activity within the fields of science and medicine which, directly and indirectly, contributed to the eventual formation of the Royal Society. By 1660 Cromwell was dead, the Stuart Restoration was under way, and Charles II was keen to improve the intellectual climate of the nation.

In 1662 the king granted the first charter transforming the 'invisible college' into the Royal Society which by a second charter less than a year later acquired its final imposing name of 'The Royal Society of London for Improving of Natural Knowledge'[16]. Another two years later appeared the first volume of the Society's *Philosophical Transactions*. The early volumes of the journal would make absorbing reading on their own account; as a record of the diverse subjects exercising the minds of the founder members of that august scientific society they are even more compelling. In those early years there appears to have been a healthy balance between papers concerned with the physical sciences and with their biological counterparts. Robert Boyle is seen to take time off from his air-pump experiments to support Richard Lower's work on transfusion of blood which graduated from animal-to-animal transfusion to an 'Experiment of Transfusion, practised upon a man in London'. The detailed report of the latter describes transfusion of blood from a young sheep through a 'silver pipe' directly into the arm of a scrupulously supervised human volunteer. We are told that the man, one Arthur Coga, not only survived the experiment but 'found himself very well' and was so enthusiastic that he asked for another transfusion a few days later. Probably fortunately for the patient, the experimenters were reluctant to comply with his request. Later the society reported experiments in Dantzick (sic) on 'Infusing Medicines into humane veines'[17].

The fifth issue of the *Transactions* was published in London on the 3rd July 1665. It concluded with an announcement to the effect that publication, and indeed all 'Publick Meetings' and 'Correpondencies' might be suspended because of the 'present Contagion'. In this year of the Great Plague in London printing operations were indeed removed to Oxford until the following

February, but publication continued with only slight delay, the journal having accomplished its own isolation strategy. Infectious diseases as such, in man or in animals, were treated in the early volumes only indirectly as reports or criticism of treatises published on the subject such as Willis' *Diatriba de Febribus*, Sydenham's *Methodus Curandi Febres*, and Nathaniel Hodges' *Historical Account of the Plague of London in 1665*. Neither Willis nor his commentators restricted their arguments to fevers; other subjects for discussion included 'nervous and nutritious juices', 'pathology of the brain', and what was referred to as 'physiological considerations of the *Soule of Brutes*'. Nevertheless, observations on animals and their diseases which found their way into the annals of the society at this time were few and casual[18].

In spite of the corroboration offered by the contemporary works of Redi, who refuted the doctrine of spontaneous generation in elegant experiments in 1668 (*cf.* chapter 3, p. 45 and n. 14), and of Leeuwenhoek's remarkable microscopical observations, the budding scientific revolution of the seventeenth century failed to make a decisive leap forward concerning the nature of contagion. Towards the end of the century there is some slight evidence that the combination of Leeuwenhoek's observations and the presence of worrying cattle epizootics in Europe were beginning to affect thinking on disease causation. It is possible that the prevailing views of Sydenham and his followers, with their emphasis on the importance of the 'epidemic constitution' of the atmosphere, still acted as an effective brake on development of theories of contagion. Sydenham, friend of Locke and of Boyle, never became a fellow of the Royal Society; but Leeuwenhoek, the self-made foreign amateur microscopist, had most of his observations published and discussed by the Society, and joined its roll of members in 1680[19].

One of the early issues of the *Philosophical Transactions* which so frequently featured one or other of Leeuwenhoek's letters, reporting observations of animalcules and other manifestations of life revealed by the lenses of his microscopes, included also pregnant comments on cattle diseases. They constituted, in the pages of the transactions of the Royal Society, a forerunner to far more extensive observations on cattle diseases which were to appear throughout Europe during the following century, when

social and economic pressures created by severe cattle epizootics
forced the authorities, and the medical profession, to search
urgently for means of controlling diseases which were decimating
the livestock populations on which they all depended. The
comments were in response to a letter received by Frederick Slare
in December 1682. Slare was a younger cousin of Theodore Haak,
a founder member of the society who had arrived from Worms in
1625. Like his more famous cousin, Slare kept up a correspondence
with acquaintances on the continent, and just before Christmas,
1682, he received a letter from a Dr Wincler, chief physician to the
Prince Palatine. Wincler reported the spread of a cattle 'murren',
from the Italian border through Switzerland to the German
provinces. He wrote of infected pastures and warned that
'...persons that carelessly managed their cattle without a due
respect to their own health, were themselves infected and dyed
away like their beasts', echoing, in less poetic language, Virgil's
earlier warnings (*cf.* p. 11). To the report in its published form
were appended the comments of Dr Slare, who by then had had an
additional report of an outbreak of what may or may not have
been the same disease (the first was probably anthrax, the second
more likely rinderpest). Slare had first been introduced to a session
of the Royal Society by Robert Hooke in 1679 in order to show
certain experiments on spermatozoa, just described by Leeuwenhoek, and had been admitted a fellow the following year, at the
same time as Leeuwenhoek. Now he wrote:

> It were worth the considering whether this Infection is no carried
> on by some volatile Insect, that is able to make only such short
> flights as may amount to such computations. For the account of the
> Ancients concerning the grand pestilential Contagions is very little
> satisfactory to this Age who derive from it a blind putrefaction,
> from the incantation of ill Men; or from the conjunctions of the
> inauspicious Planets. I wish Mr. Leeuwenhoek had been present at
> some of the dissections of these infected Animals, *I am perswaded*
> *He would have discovered some strange Insect or other in them*[20].
>
> (my italics)

In spite of this perceptive comment Slare did not further pursue the
subject but turned to chemical studies more in line with the
prevailing preoccupations of the Royal Society at the time. Thirty
years were to elapse before the possibility of a connection between

Leeuwenhoek's animalcules and cattle epizootics was to surface again, when cattle plague was rampant in Italy.

Whereas in England the 1660s belong to the Stuart Restoration and are associated with Charles II's favourable attitude to the arts and the sciences, in France the decade was also one of change, but of a somewhat different complexion. In 1661 Mazarin died, leaving Louis XIV at the age of 23 to shape the lifestyle of himself and of the nation to his own taste. His style was no less flamboyant and even more frivolous than that of his English counterpart, and he was infinitely more interested in the arts and architecture than in any scientific subject. The 1660s nevertheless brought a notable by-product of the intense preoccupation with horsemanship which flowed freely in an age focused on the glittering court of the Sun King and the wars he fought. In 1664 was published at Paris the first edition of *Le Parfait Maréschal*, the work of one Jacques Labessie de Solleysel (1617–80), who held a senior position in the Royal Riding Academy in Paris. After early studies at Lyons he had been drawn irresistibly to equestrian pursuits. Eventually he became Master of the Horse to the French Ambassador, negotiating the Peace of Westphalia at Münster in 1645, and during his prolonged stay in Germany he closely observed local veterinary practices. Upon his return to France he joined the Paris riding academy and devoted the rest of his life to it. The *Parfait Maréschal* is a testament to his common-sense and to the intelligent way in which he translated his experiences in the field and in the riding academy, at home and abroad, into a general guide for those concerned with the welfare of horses. Nowhere in his book is this more evident than in the sections dealing with infections. He understood very well the problems of contagion and the value of isolation, and wrote in his account of glanders:

> ...not only it communicates its Venom at a small distance, but it infects the very Air, and seizes on all the Horses that are under the same Roof with him that languishes under it. And therefore as soon as you perceive the least sign of Glanders, you must separate the sick Horse from all his Companions, and not suffer him to drink out of the same pail with 'em;...[21]

His preventive measures were limited to isolation, and there was no suggestion of destruction of animals as valuable as horses; all the more since distinction between glanders and the less severe farcy was still uncertain, and transmission to man unheard of. Solleysel was much preoccupied with the influence of phases of the moon on all aspects of veterinary practice, and like many before and since he advocated a most powerful and infallible remedy against rabies in man and in all species of animals. Having listed the unremarkable ingredients he informed his readers that the same potion might be used against the plague, and in general for those who 'breathe infected air'. It was another example of the desperate search for protection against dreaded infections in man and in animals which were to resist treatment for centuries to come; and yet certain empirical measures, with their roots in folk medicine, were to have some impact in the following century.

CHAPTER 3

The impact of cattle plague in the early eighteenth century

There were a number of indications in the latter decades of the seventeenth century of changing attitudes and of more realistic approaches to the problems of the nature and spread, and hence to the eventual conquest of, infectious disease in epidemic or epizootic form. At the same time, from about the year 1700, there came also a greater awareness of results from abroad, and references to the works of others became more common in papers and treatises. This may to some extent have been a function of a more widely based exchange of thoughts and viewpoints through the medium of the Royal Society and its *Philosophical Transactions*, and similar societies and journals in other European countries.

In addition to such developments facilitating the flow of ideas within the continental states and across the English Channel, a comparative approach was furthered and consolidated in the eighteenth century by two natural protagonists. One was a disease of man, the other one of cattle: smallpox and cattle plague, respectively. Sydenham's work on smallpox in the latter half of the seventeenth century[1] serves as a reminder that the disease in the 1660s and 1670s was causing extensive and increasingly severe epidemics in England and on the European continent. No respecter of rank, smallpox made such inroads among the legitimate Stuart heirs (as it did also in other royal families in Europe) that it paved the way for the Hanoverian succession in England[2]. If not quite on the same scale demographically, but certainly psychologically, smallpox towards the end of the seventeenth century was gradually replacing bubonic plague as the main and most obvious threat to man in European epidemics. At the same time there is evidence of the emergence of the practice of 'transference', i.e. attempts to 'transfer' smallpox (or plague or syphilis) from human sufferers to animals or even to inanimate objects. Any connection between such practices and the origins and specificity of the many distinct animal poxviruses must remain highly speculative. In rural areas throughout Europe there existed also in the seventeenth century the custom of 'buying crusts' from the smallpox lesions of mild cases in order to produce subsequent immunity by voluntary submission to what it was hoped would be mild cases of the disease. The latter practice may have had its roots in the age-old Indian and Chinese customs of 'snuffing' smallpox crusts with the same end in view[3].

Quite how these various practices evolved and eventually merged we do not know. Developments may have been wholly separate and independent; but in 1701 one Giacomo Pylarino, a Greek educated at Padua, was in Constantinople during a serious outbreak of smallpox, and saw a Greek woman performing what he called 'transference'. She inoculated matter from pustules from a mild case in an operation which subsequently became known as inoculation or grafting. The term 'inoculation' was introduced in 1714 in a letter to the Royal Society by another colourful character, Emanuel Timone, born on a Greek island of Italian parents[4]. Although the terminology was different, the practices described by Pylarino and by Timone were identical and derived from near-Eastern traditions; Timone mentioned sources in Circassia, Georgia, around 1674. Following the publication of Timone's account, and of a small volume on the same theme by Pylarino in 1715[5], knowledge of the method slowly spread throughout Europe and the newly established colonies on the other side of the Atlantic. In spite of the helpful publicity engineered by various factions of society led by Lady Mary Wortley Montague, Voltaire, and the Princess of Wales[6], the practice of inoculation continued to be disseminated only slowly, and only within limited circles. With the help of Sir Hans Sloane, who for a number of years during the 1720s and 1730s combined the presidencies of the Royal Society and of the College of Physicians, these circles came to include the Royal Society and eventually also the majority of the medical establishment. Although inoculation could not be said ever to have become generally accepted by the population at large, some successful practitioners achieved wide acclaim for their specialised service. Later in the century attempts were made to develop similar protective measures against what were considered similar diseases: measles in man, and rinderpest in cattle[7]. In the 1790s, Italian veterinary authorities suggested that the 'variola of turkeys' was related to similar diseases in quadrupeds, and some believed that turkey-pox could be transmitted to sheep. During one outbreak, unsuccessful attempts were made to immunise such birds[8].

It was no accident that rinderpest came to be included in such attempts. As far as veterinary medicine was concerned, the eighteenth century had begun as the seventeenth had ended. The discipline as such was non-existent. Although here as in many other areas the century of enlightenment was to bring changes for the better, they were slow in coming. To a considerable extent they developed through sheer necessity. As human populations recovered from the plague years, they began building up their numbers of livestock; then, only a decade into the new century, disaster struck. In 1711, rinderpest (or cattle plague as it was then known) reached Italy from Dalmatia, and soon became established throughout Europe and from 1714 in the British Isles as well. Although the initial outbreaks in Italy and in Britain had been successfully contained by 1715, the disease spread and became endemic elsewhere in Europe. From such areas it was periodically re-introduced until almost the end of the century. At various times, and in various places, foot-and-mouth disease and infectious bovine pleuropneumonia were present simultaneously and gave rise to diagnostic confusion. Between them the three diseases, but rinderpest in particular, presented a continuous threat to cattle herds and hence to their owners and communities in general. During periods of epizootics among livestock, famine and hardship were never far away, and this is amply documented in the writings of the time. Numerous works on cattle diseases published during the eighteenth century gave graphic descriptions of the devastation and general misery caused to man and his herds, in Britain and elsewhere in Europe[9].

At the beginning of the eighteenth century, the continual presence of such threats in epizootic form to cattle herds forced whole communities, and in the absence of trained veterinarians the medical authorities, to concern themselves with animal disease in a way which had been unheard of in previous centuries. By mid-century the need for a new profession of trained veterinarians had become acutely felt and painfully obvious. In a Europe ravaged by cattle diseases, farmers and herdsmen could look for help only to itinerant, self-taught or untaught 'cow doctors' or 'cow leeches' whose advice and 'treatment' were at best useless, at worst actually harmful if not lethal (Fig. 6). Nevertheless, when the first veterinary school opened its doors to students at Lyons, France, in February, 1762, its teaching, no less than previous literature, placed a disproportionate amount of emphasis on the equine species, in spite of verbal acknowledgement of the ubiquitous threat of cattle epizootics. For this and other reasons both the Lyons school and others soon to follow in France and elsewhere had to overcome manifold difficulties before they became firmly established (*cf.* chapter 5).

In 1700, that was well in the future; but the problems of livestock epizootics were moving uncomfortably close. Slare's reflections on outbreaks of anthrax and perhaps some form of cattle plague in the 1680s were discussed in the previous chapter (p. 31). Throughout the seventeenth century rinderpest, often accompanied by and confused with, foot-and-mouth disease and bovine pleuropneumonia, had been intermittently imported from Russia via Hungary to Italy with cattle transports from Dalmatia. Although such outbreaks frequently spread further into Europe they had usually been contained before reaching alarming proportions; in addition, they had been overshadowed by, even when immediately preceding, the great epidemics of plague in man.

In 1709 a new wave of rinderpest began to travel from Tartary through Russia to Poland and Dalmatia, and in 1711 it reached northern Italy with an infected herd in transit from Dalmatia[10]. The accepted contemporary version, repeated by all the highly respected authors who were to comment on the outbreak within a short time insisted that one, and only one, straying, infected ox initiated the spread of infection. In retrospect this may seem open to doubt. In a herd under stress, moving long distances through unfamiliar country, with perhaps inadequate grazing on the way, an infection such as rinderpest would spread rapidly. There would

Fig. 6. The Cow-Doctor: more impressive in appearance than in healing powers. Engraving by Cousens after Tschaggeny (courtesy Wellcome Institute Library, London).

be more than one opportunity for infected individuals to pass on the contagion to native beasts along the road. But apocryphal or not, the story of the one ox, accepted as fact at the time, was a useful model for developing epidemiology: the idea of a distinct focus from which radiated further spread[11]. It was also a warning and a lesson to be learned in other parts of the country where there was immediate awareness of the threat posed by the importation in the north. For once it had taken hold, rinderpest spread with alarming speed.

In the event there was no dearth of suitable host animals, and their keepers were unprepared for such a virulent and extensive epizootic. By October 1711, within a few months of its onset, rinderpest affected all of the Venetian territories and showed no sign of abating. Alarmed, the Venetian Senate appealed to the region's highest medical authority, the medical faculty of the University of Padua. Its *Professor Primarius* then was Bernardo Ramazzini (1633–1714), who in 1700 had published the first edition of his classic volume on the diseases of workers. He had always been concerned with the epidemiology of his times, and had indeed written extensively on the effects of weather on the health of the people, both directly and indirectly. In the 1690s he had pointed out the deleterious effects of unseasonal weather and heavy rains on crops, in turn affecting first domestic animals and then man. Acknowledging the nature of the prevailing rusts and mildew of the crops as 'contagious blights', Ramazzini felt that the '...cause of death in animals ['variolous eruptions' in sheep, 'suffocation' in pigs]'...arose from the acid nature of the mildew; and in addition to the 'morbid constitution of the atmosphere...noxious to beasts' he blamed contamination of forage for the blood becoming acid and circulating 'feebly, and either whole flocks of sheep died suddenly, or were seized with small-pox'. A thorough examination of the pustules convinced him that they were identical with smallpox pustules in children. Diseased plants also affected bees, the main source of sugar; finding 'no sweetness from the calyces of flowers, but a bitter poison, [they] either died or migrated'[12].

Now in 1711 approaching his eightieth birthday, almost blind and plagued by ill-health, Ramazzini nevertheless responded immediately and with admirable good sense to the challenge of rinderpest. He devoted his thirteenth annual oration to faculty and students at Padua in November, 1711, to a discussion of the outbreak and its possible causes, and to recommendations for therapy and preventive measures. The latter were little different from other attempts to control epidemics and epizootics by means of simple (or perhaps in the context of the times, not so simple) cleanliness, isolation of affected animals, and fumigation of stables; but they were rational and clearly stated. In his view of the nature of the contagion, Ramazzini again came down firmly on the side of common sense. He rejected astrological explanations, so popular in previous centuries, and presented a theory not unlike that of Fracastoro, believing that infections were caused by 'seeds' of disease, which lodged in susceptible organisms where they multiplied and produced morbid conditions. As in the work of Fracastoro, there was no suggestion that such 'seeds' might be animate. His concluding plea for heavenly guidance may appear incongruous today in the light of his lucid reasoning; it was quite in keeping with the outlook expected in his times[13].

Towards the end of his life, then, Ramazzini had a firm grip on the basic elements of epidemiology. His experiences in the last decades of the seventeenth century behind him, he had devoted his inaugural address as *Primarius* at Padua in 1709 to a discussion of the possible effects of the intense cold of the first three months of the year, warning that they might give rise to severe summer epidemics of acute fevers with high mortality, and promising to lecture that term on such fevers. His last annual oration, in 1713, was concerned with an epidemic of plague at Vienna in which the victims were mostly poor inhabitants of the suburbs. However, in 1711 he chose to compare cattle plague not with bubonic plague in man but with smallpox, drawing attention to the similarity between the development of pustules after five to six days, the high mortality, and the prevalence of deaths between the fifth and the seventh day. Comparisons of this sort were taken up by subsequent authors and later in the century led to attempts to use inoculation as a means of control; they appeared in many later treatises on cattle plague. As the disease continued its progress in Italy and elsewhere in Europe and in the British Isles, there were to be many such treatises. Until the advent of the veterinary schools, from 1762 onwards, they were all written by physicians or surgeons, most of them prominent ones. The majority of these tracts also share another theme introduced by Ramazzini: reproach directed at medical colleagues who found it beneath their dignity to discuss cattle and cattle diseases, no doubt fearing to be associated in the minds of their patients with the despised cow leeches.

First among the medical establishment in Italy to join Ramazzini in his impassioned plea for adequate measures to be taken against the rapidly spreading threat of the cattle plague was Giovanni Maria Lancisi (1654–1720) (Fig. 7), personal physician to successive popes. His employer at this time was Clement XI, whose Papal reign began in 1700, included support for the Old Pretender in 1715, and ended with his death in 1721. At the time of the outbreak Lancisi was at the height of his powers, and his work on cattle plague was a natural extension of his interest in the epidemiology of influenza and of malaria which followed his major treatises on cardiac disease[14]. It was also born of necessity: an attempt to prevent the epizootic in the north from reaching the Papal states. Although in this he was not initially successful, the policy of quarantine, isolation and slaughter which he recommended and vigorously enforced did mean that the disease was brought under control within a relatively short period. By the time Lancisi wrote his account of the epizootic it was from the comfortable position of a man who had stemmed the tide of disaster, albeit with the help and backing of the awesome machinery of the Papal state. There are chilling reminders in the text of the use and extent of its powers, e.g. '…merchants who had by evil deceit broken the ban on business were thrown into gaol to be suitably punished in due course (deceitful merchants put in chains)'. They had certainly been warned. Edicts had been issued, warning of the impending danger and carefully describing the precautions which farmers, merchants, everyone concerned with the care of domestic animals, should, and indeed were obliged to, undertake. Lancisi included copies of all the edicts in his text. Unlike the scholarly contents of the book, written in Latin, the edicts are in the Italian vernacular, presumably to facilitate understanding by the public. Both in the edicts and in the general directives described by Lancisi there was far more emphasis on practical measures than on any 'beseeching of Divine Help', although a few edicts are concerned with suitable prayers for deliverance, the ultimate to be recited 'at the time of the Sacred Advent, at the sound of a trumpet…'[15].

Fig. 7. Giovanni Maria Lancisi (1654–1720); engraving by Marcucci after Cleter (courtesy Wellcome Institute Library, London).

Lancisi's book is much more than a self-congratulatory document. He was not a man to rest lightly on his position as Papal physician or on his own success in containing the outbreak. Like Ramazzini he realised that there were valuable lessons in epidemiology to be learned from the study of animal disease, and he took pains to discuss and analyse those lessons at length. The conclusions he reached were to remain as guidelines for several generations of physicians and, after 1765, veterinarians, who between them struggled to control the countless epidemics and epizootics which formed an integral part of the problems of existence in the eighteenth century and beyond.

Having been instrumental in controlling the outbreak, Lancisi proceeded to review the situation and to attempt to apply the knowledge he had gained as a basis for a general theory of the nature of infections. In order to do so he summed up past and present knowledge with particular reference to the question of the existence of a *contagium animatum*, going back as far as Varro and Columella and their views on swamp fevers, a subject he was to treat at greater length in a separate treatise[16]. At the other end of the spectrum, he considered ideas put forward only two years earlier by his compatriot Carlo Francesco Cogrossi (1682–1769) in letters to Antonio Vallisnieri (1661–1730). From this background Lancisi proceeded with caution to conclude that the facts of transmission, directly from beast to beast or less directly carried via herdsmen and farmers, and their dogs, showed beyond reasonable doubt that the disease spread by means of specific seeds (*semina*) of cattle plague. In contrast to Cogrossi and Vallisnieri, Lancisi could not accept a theory of *contagium animatum*. Instead, he saw the seeds of disease as 'tenuous corpuscles of the character of a particular poison' or 'pestiferous ferment' which in the form of 'molecules' passed from beast to beast. Having once invaded a new host organism these inanimate entities were there able to multiply. Lancisi's ideas, accepted and repeated by most subsequent authors in the eighteenth century, accidentally had elements of current concepts of viral transmission and replication while missing the basic premises necessary for a bacteriological interpretation of the spread of infectious diseases. A much closer approach to the latter was achieved by Cogrossi.

At the time of the outbreak in northern Italy, Cogrossi was a young practising physician in his home town of Crema in Lombardy, with none of the well-established eminence of a Ramazzini or a Lancisi. A few years earlier he had been taught at the University of Padua by Ramazzini and by Vallisnieri. In 1713 he was appalled by the devastation he saw all around him and what he called the 'funereal traces' of the great slaughter produced by 'this savage epidemic'. Prevented from making postmortem examinations by the justified haste with which it was deemed necessary to bury the carcases, and not otherwise involved in a practical way in the control of the outbreak, Cogrossi set down his purely theoretical thoughts on the nature of the disease in a letter to his erstwhile teacher Vallisnieri.

After the briefly voiced suggestions of the existence of animate disease agents by Slare and others, and the lengthy but problematic exposition by Kircher in *Scrutinium pestis*, Cogrossi's *Pensieri*[17] presents the first clearly reasoned and thoroughly argued defence of the theory of *contagium animatum*. Writing in 1713, Cogrossi was also able to draw on the analogy provided by Redi and by two of his associates in 1687. The two publications provided the first appearance of simultaneous reflections on the parallel problems of spontaneous generation and living disease agents which were to be conclusively resolved only in the nineteenth century. When Francesco Redi in 1687 was able to refute the theory of spontaneous generation of maggots in decaying meat, his friends G. C. Bonomo and G. Cestoni in the same year could provide the first complete evidence for the causal role played by the *acarus* mite in scabies in man[18]. It should, however, be noted in this connection that Cestoni did not approve of this application of his results; indeed, he pronounced himself very disappointed that Cogrossi should make the comparison. His own observations of the mite were, he said, 'true philosophy' while the speculations of Cogrossi were effectively 'theology'.

Taking the *acarus* evidence as his main source of inspiration, Cogrossi had postulated, in his letter to Vallisnieri, that similar processes involving organisms smaller than the *acarus*, of a size to be seen only with the help of Leeuwenhoek's microscopes, might be responsible for the development and spread of infectious diseases: not only the present cattle plague, but also numbers of other transmissible diseases of man, animals and plants, from syphilis to corn rust. And indeed, until the end of the nineteenth century, the pathogenesis of scabies was to remain a favourite paradigm used for reference and comparison by very many physicians and veterinarians and even botanists interested in the developing sciences of parasitology, microbiology and bacteriology[19].

In spite of their differences in interpretation, the above authors shared a belief in the advantages to be gained from study of animal disease, and a healthy disregard for the opinions of those who considered it undignified to concern oneself with animals lower than man. Ramazzini had explained that discussion of cattle diseases need not be beneath the dignity of medical men. Lancisi berated 'delicately fastidious' physicians who found it 'inconsistent with their image and dignity to compromise their minds by attention to that department of medicine which is called Veterinary, and which serves the health of Beasts'. He explained that his own attitude was due not just to the realisation of the extent to which man was dependent on animals for his own welfare, but also to a 'deep sense of emotion towards living brutes and the whole range of creatures from the separation of whose innards by the scalpel has flowed and almost been perfected that art of surgery which, because we progress from examining entrails of other living things to knowledge of human entrails, they call "comparative"'. Quoting several instances of Hippocratic uses of comparisons with goats, sheep and cattle in exploration of human disease, Lancisi concluded that anybody reproaching doctors for concerning themselves with the causes and control of the present pestilence would be dissociating themselves from all that was 'true, useful and honourable'[20].

Thus the first Italian outbreak of cattle plague in the eighteenth century was brought under control on the advice of two of the most renowned physicians of the day, one of them personal physician to the Pope and both of them prepared to throw all their social conscience and all their medical knowledge and experience behind an all-out campaign against a threatening cattle disease. In so doing they took important steps to open up the field of comparative epidemiology and pathology. Both on compassionate and on practical economic grounds were there compelling reasons for a fresh approach to animal care in the century of the Enlightenment.

Shortly after the first wave of the great cattle plagues of the eighteenth century had been brought under control, there was a severe outbreak of bubonic plague in the South of France, radiating from Marseilles. With the extensive seventeenth century outbreaks still fresh in many minds, apprehension at the thought of further spread was immediate and gave rise to a number of short works concerned with plague, its nature, spread and prevention, published in France and in England between 1720 and 1722. One short book was by Richard Bradley, later (from 1724 to his death in 1732) professor of botany in the University of Cambridge. Bradley probably had no formal medical qualifications[21] but his flair for a comparative approach to problems of contagion must remain undisputed. His pamphlet on the plague at Marseilles, 'published for the preservation of the people of Great Britain', appeared in 1721, a year which also saw the publication of his *Philosophical Account of the Works of Nature*; during the five years between 1717 and 1722 he published a number of works concerned with gardening and husbandry, and advice about prevention of infection. In all of these writings Bradley took as his point of departure the diseases of plants he first wrote on in 1718, in a chapter on 'Blights'; from there he extended his findings and conclusions to diseases of animals and man. Unlike his contemporary Richard Mead (1673–1754) he had no time for the prevailing notion that contagion was spread by means of corrupt air, and that disease agents were 'substances in the nature of a salt, generated chiefly from the corruption of a human body'[22].

On the contrary Bradley believed that blights of plants, and epidemics in man and in animals, were caused by animalcules (and like contemporaries from Slare to Cogrossi and Lancisi he sometimes referred to them as 'insects' or 'worms'). In his pamphlet on the plague at Marseilles he drew comparisons not only with the Italian and English outbreaks of cattle plague in 1711–14 but also with the previous 'murren' reported in 1682 by Wincler and by Slare (*cf.* p. 31). He also referred to reports in the *Philosophical Transactions* of the Royal Society (Bradley became an F.R.S. in 1712) concerning blights of trees spread by 'Flies', remarking that since large insects such as locusts and big flies are carried long distances by winds, the 'smaller kinds, which are lighter than air' may be moved great distances by winds, thus helping to spread disease. Elsewhere he used the scabies analogue mentioned above to illustrate the virtue of cleanliness, mentioning also the danger of transmitting both scabies and other 'distempers' by the use of contaminated towels and other 'Linnen'. His belief in the existence of a *contagium animatum* was shared by his contemporary Benjamin Marten, who was mainly concerned with tuberculosis in man, and by the French physician Jean-Baptiste Goiffon (1658–1730), who wrote on the outbreak of plague at Marseilles in his capacity of someone responsible for preventing the plague reaching Lyons. Goiffon particularly emphasised that agents of plague and of other transmissible diseases must necessarily be capable of biological reproduction in order to create sufficient numbers of their species to sustain the rapid spread from person to person observed in epidemics. He had earlier, at official request, written a tract on the spread and control of cattle disease in France in 1714. Unfortunately his views on epizootics are lost to us since no copies of this book have been found to have survived, even in France[23].

Benjamin Marten's *A new theory of consumptions* does survive, albeit in a limited number of copies. In the early years of the twentieth century Charles Singer, who possessed one of the existing copies, revived interest in Marten by reprinting the greater part of Marten's chapter 2, which contains his remarks on agents of transmissible diseases, tuberculosis in particular,. but also smallpox and plague and 'several other distempers that we find at some times not only to afflict man but to destroy beasts;...'. Such diseases of man and of animals might be caused, Marten wrote, by '*Animalcula* or wonderfully minute living Creatures' of different shapes and sizes, too small to be seen with the naked eye, but all 'produced from an *Ovum* or Egg'. Singer's hyperbole dubbed Marten 'a meteoric prophet of the parasitic nature of the infectious diseases'[24]. In his carefully chosen extracts he did include Marten's reference to Slare's comments in the *Philosophical Transactions*; but he omitted the numerous references to and lengthy *verbatim* passages from the work of the 'so often quoted Dr. Andry'. Yet these clearly show the seminal influence of this author on Marten's 'armchair philosophy'. Consideration of both their works, and others referred to by Marten, serves to demonstrate that the concept of a *contagium animatum* at the beginning of the eighteenth century was not as rare as is sometimes assumed. The writings of Nicolas Andry (1658–1742), unkindly called *Homo vermiculosus* by a critic, also offer some insight into the use of terms such as 'insects' and 'worms' as equivalents of *animalcules* as well as of much larger insects and even reptiles at this time.

Andry began a chapter on the breeding of worms in human bodies, with a definition of 'what a worm is':

> Since Worms are included in the *Genus* or *Kind of Insects*, it is proper, in order to understand what a worm is, to explain first the Nature of an Insect. An Insect then is a compleat Animal, distinguished or divided by several Incisions in Form of Rings and Cycles, by means of which it breaths, and by reason of which it is called an *Insect*. Such are the Scorpion, the Ant, the Fly...[25]

This was a representative view in the decades before the publication of Linnaeus' *Systema naturae* in 1735. Both Marten and Andry referred to advances made by use of microscopes, and hopes of further discoveries as microscopes were improved.

In 1730 Thomas Fuller published *Exanthematologia*, a volume of detailed clinical descriptions of the various 'eruptive fevers', with some general considerations on their nature. Writing in the seclusion of a country practice in Kent, with few opportunities to discuss his ideas with colleagues, Fuller also came to some astute conclusions. He believed that the various exanthemata were caused by 'virose corpuscles'; he emphasised their specificity, the fact that 'every Sort of venomous Fever is produced by its proper and peculiar Species of Virus'. In a century when Rhazes' ideas of smallpox as the inevitable outcome of a necessary 'fermentation' of an inherent quality of the blood (*cf.* p. 17) had been transposed into a belief in the 'innate seed' of such diseases[26], Fuller attempted his own version in an imaginative synthesis of ideas. He added the notion of an external '*afflatus genitalis*' required to combine with specific '*ovula*' of the blood in order to produce the specific disease.

Fuller freely used broad comparisons to express his thoughts, writing of specificity that '...the Pestilence can never breed the Small-Pox, nor the Small-Pox the Measles, nor they the Crystals or Chicken-Pox, any more than a Hen can a Duck, a Wolf a Sheep, or a Thistle Figs; and consequently, one Sort cannot be a Preservative against any other Sort'. He was at pains to point out his disagreement with the suggestions of certain other authors that one smallpox-like but benign eruptive complaint of children might be identical with the *Grando Porcorum*, an eruptive disease of hogs. He wrote: 'I confess these Pustules in Children are not much more like the *Grando* in Hogs, than the Constellation called the Celestial Bear is like the Terrestrial Beast: but Custom is a Tyrant for Names as well as Things, and will have it so'[27]. He did not believe in mincing words, and from his country practice still upheld the superiority of his profession over lesser breeds. He dedicated the book to Hans Sloane, and in an introductory historical passage he told Sloane of the early practitioners that 'Most of these were only Surgeons, and understood little, but stanching of Blood, and healing of Wounds and Ulcers'. This was on the eve of the ascendancy of the Hunters; and fifteen years before Parliament passed the Bill dividing the Company of Barber-Surgeons into two separate institutions[28].

CHAPTER 4

Cattle plague in England and on the European continent 1714–80

As seen in the previous chapter, the first well documented epizootic of rinderpest began in Italy in 1711 and within a few years enveloped all of the European continent. It followed a classical progression seen in many epidemics and epizootics before and since. From the Middle and Near East it invaded Italy via Dalmatia, and then moved on to Austria, the German States, France, and the Low Countries. Once near the Channel ports, such outbreaks were poised for entry into the British Isles[1].

In the middle of July, 1714, the disease appeared in herds on three farms at Islington, then one of the rural areas bordering on expanding London. The authorities reacted with commendable, in the circumstances even surprising, speed. It was a time of political upheaval; Queen Anne died on the first of August, and George I did not arrive in London until mid-September, speaking no English and knowing little about the administration of the country. The story of the successful control of the invading cattle plague of 1714 is a testimony to the efficiency of the government of the time, irrespective of changing political fortunes and absentee monarchs[2]. In Rome the Pope had appealed to his personal physician; the English Lord Justices put the matter into the hands of a surgeon to the Royal Household, Thomas Bates (d. 1760), and the Lord Chancellor ordered four Middlesex Justices of the Peace to assist his enquiries. The Pope had been fortunate in the services of Lancisi, and Bates proved an equally effective choice for London.

It is not known whether Bates knew of Lancisi's campaign; he certainly could have had no access to his account of the Italian outbreak, which was not published until the following year. Before

1709 Bates had served with a Navy regiment in the Mediterranean, and may have had contacts in Italy; and an extract of Ramazzini's less extensive report had been published in the *Philosophical Transactions* in March, 1714[3]. On the other hand, the mixture of common sense and received knowledge of control, which, as Lancisi pointed out, had been available since the time of Columella, could well have been Bates's own contribution. His firm and successful intervention may be seen as an illustration of another source of informed aid available when dealing with animal disease before the advent of trained veterinarians. Because of the close proximity of horses and men in times of war, army surgeons, especially those in mounted regiments, often found themselves with responsibility for the care of horses as well.

There are a number of examples of army surgeons having contributed to the literature on diseases of animals, especially horses, before the existence of veterinarians trained in veterinary schools. A contemporary of Bates, the surgeon William Gibson (1680?–1750) served for a time with the 16th Dragoons. He acquired a certain amount of knowledge of the treatment of diseases and injuries of horses during his service. As a result he wrote several volumes of advice on their care over a period of 30 years, beginning tentatively but gaining in confidence. His final treatise on diseases of horses was published posthumously, containing much clinical observation and common sense advice[4]. Bates himself had served with the Navy, presumably with no access to animals; but when he left the Navy after five years' service in the Mediterranean he had established his epidemiological competence by writing a manual on the nature and control of the fevers attacking sailors during the hot Mediterranean summers[5].

In London in the summer of 1714, with the full cooperation of the local Justices of the Peace and the Lord Justices, Bates acted swiftly, with radical measures which proved to be eminently successful. He recommended immediate isolation of infected herds, destruction of affected cows, and burning of their carcasses, sparing a few sick individuals for long enough to determine whether or not the disease was indeed specific to cattle. Aware of the social consequences and of the need to encourage recognition of the presence of the disease, he also recommended that compensation of 40 shillings for every cow destroyed should be payable to the owner. Initially accepted by the authorities, this sensible proposal was later brought into jeopardy when it was suspected that certain farmers failed to comply with the preconditions of early notification of the presence of the disease and of refraining from selling suspect cattle and their milk and meat at distant unsuspecting markets. However, the over-all policy worked in spite of difficulties, and by the end of the year the outbreak was over[6].

After 1714 England remained free from cattle plague for a period of thirty years, although the disease continued its presence elsewhere on the European continent. In the Netherlands it became endemic quite early in the century, largely owing to rejection of radical slaughter and isolation policies and to a misplaced trust in spurious 'cures' and 'medicines', some of which were described in the *Philosophical Transactions* as early as 1714[7]. As a result, what was then the Austrian Low Countries remained a reservoir for reintroduction of rinderpest to its neighbours until almost the end of the century.

In 1745 the disease again assumed alarming proportions near French and Belgian ports. By this time the *Gentleman's Magazine* and its rival the *London Magazine* had begun publishing, in 1731 and 1732, and their contents for the years 1744 and 1745 bear witness to an awareness of the danger threatening to invade the country. In November, 1744, the *Gentleman's Magazine* carried a report warning of a 'murrain' among the black cattle in Flanders; a year later Thomas Bates, now living in retirement in Hampshire, felt the need to communicate to the magazine's readers his recommendations for precautions to be taken 'in the present incurable, and contagious distemper among the cows'[8]. It had arrived in April, 1745, when a farmer at Poplar, hoping to improve his stock, bought two white calves in Holland and brought them home to 'mix the breed'. In 1711, in Italy, one Dalmatian ox had been blamed for the singlehanded introduction of cattle plague; now, in 1745, the two white calves from the Netherlands travelling to Poplar figured prominently in contemporary reports as the focus of a fresh and more serious outbreak which proved far harder to contain than the previous one in England[9], although infected hides from Flanders may also have played a part.

The 1714 outbreak in and around London had weathered the

political uncertainties ensuing from the delay over the succession
and the arrival of George I. More unsettling in the country in 1745
was the Jacobite uprising with its movements of troops and the
political ramifications of the War of the Austrian Succession. In
addition, climatic factors and possibly feeding difficulties may
have left the cattle with lowered resistance, and the strain of virus
involved may have been more virulent in 1745 than the one
responsible in 1711–14. The editor of *Gentleman's Magazine*,
commenting on a letter from a reader in December, 1745, blamed
possible 'fatal delays' for what was to follow. There was, he said,
very little reason to look for helpful vigilance to a government
which had 'for many years' been characterised by the bureaucratic
slowness of its appointed boards[10]. Whatever the reason, the
outbreak in England lasted for more than ten years, and caused
considerable hardship which is reflected both in the above-
mentioned magazines and in other literature at home and abroad.
The very high mortality was a prominent feature of the infection.
Layard in 1758 concluded that the rates of mortality in France and
in England were identical, referring to statistics provided by the
Marquis de Courtivron at the Académie des Sciences in 1748; in
one herd the Marquis registered a mortality of 91 per cent. Only
7 out of a total of 192 head of cattle in one herd recovered from
the infection, and 9 appeared to be immune[11]. In the British Isles,
this second outbreak was also far more widely distributed than
had been the case in 1714, when only counties bordering on the
London area were affected except for a few cases in Norfolk,
Suffolk and Herefordshire, which were contained within a very
short time towards the end of the outbreak, following stringent
measures of isolation and deep burying of carcasses. It is also
possible that the situation could have been aggravated by the
simultaneous presence of contagious bovine pleuropneumonia in
the 1740s and 1750s. Some reports emphasise lung involvement to
the exclusion of other symptoms. On the other hand, the high
mortality and the generally observed rapid progress and conclusion
of individual cases point strongly to rinderpest in most descriptions.

The pages of the two London magazines, and of their Scottish
counterpart, the *Scots Magazine*, as well as private letters of the
period, testify both to the extensive presence of the disease and to

the hardship it caused throughout England and Scotland. Reports exist from the London area and its bordering counties, west into Somerset and Wiltshire and east into Essex and Suffolk and north through Worcester, Derbyshire, Nottinghamshire and Yorkshire to Durham and Argyll. Stories of farmers who lost all their cattle in spite of desperate efforts to save them are legion, although others, unpredictably, were spared in spite of being in areas visited by the disease[12]. The latter inevitably gave support to the growing movement in favour of attempts to inoculate herds with the disease in well-controlled circumstances, in the manner of the smallpox inoculation which at this time was gaining ground as a possible means of controlling and mitigating outbreaks of smallpox in human populations.

The mid-eighteenth century outbreak of cattle plague lasted for more than ten years, beginning in April, 1745, and releasing its stranglehold on British cattle farming only towards the end of 1757. Even then it manifested its presence in sporadic minor outbreaks in various parts of the country for another ten years. The length and severity of the epizootic are reflected in contemporary medical literature as well as in the aforementioned more popular magazines. Already in 1745 members of the medical profession were aware of the danger and began to contribute observations and advice both in and out of the forum of the Royal Society. By now it appears that the need for intervention was so obvious that few physicians felt the need to apologise for an interest in animal disease. The Secretary of the Royal Society, Cromwell Mortimer, M.D. (*d.* 1752) wrote in November, 1745, that he felt it his duty as a physician to do what he could to alleviate 'this publick Calamity', and that consequently he had sought information by 'visiting several of the Cow-houses near the Skirts of *Westminster* and *London*' (of particular concern was the milk from a herd in 'the Vineyard in *St. James's Park*'). There were still echoes of former attitudes, however, in his subsequent remarks on the need to lend compassionate assistance to members of the 'brute Creation'. In view of the high prices fetched by horses and dogs, he wondered aloud at the failure of their owners to encourage 'Gentlemen of higher Degrees of Learning than the Farrier and the Cowleech' to take an interest in their diseases[13].

Earlier in the same year Theophilus Lobb, M.D. and F.R.S.

(1678–1763), had included in a collection of letters on the plague and contagions in general one on 'contagious diseases among cattle'. It was addressed to John Milner, who as Justice of the Peace had been in charge of the administrative side of control of the 1714 outbreak, but there was no reference to Thomas Bates, the surgeon who had masterminded the measures of control. Lobb, of non-conformist stock and originally trained for the ministry, had catholic tastes and interests. In middle life he had absorbed enough medical knowledge from friends and neighbours to earn a degree of M.D. at Glasgow in 1722, and later to become a licentiate of the College of Physicians. Earlier he had written a volume on smallpox, with practical advice and case histories, many collected during his ministry at Yeovil, Somerset, where there had been 'no physician besides myself' during a local outbreak in 1717–18[14]. Having become a Fellow of the Royal Society in 1729, he abandoned the ministry and devoted himself wholly to medical practice at the age of 58.

In the early 1740s Lobb wrote a number of letters on bubonic plague to the then President of the Royal Society, Martin Folkes. There was no immediate threat of reintroduction, but the Marseilles outbreak of 1720 was still fresh in the memory. Lobb mentioned in his foreword that quarantine regulations were still in force, as an indication of continuing danger. He also emphasised that even in the absence of plague his suggestions could be useful in cases of 'other infectious diseases' such as occurred in 'Ships, Jails, and other Places where People are crowded together'. He finally spared a thought for the doctors and nurses who cared for victims in serious epidemics, and suggested protective measures. The letters are full of practical suggestions for control, fumigation procedures, and 'medicines'. They were collected in a volume dedicated to the Speaker of the House of Commons, with an equal number of letters to an anonymous recipient, most of which are in a much less scientific vein and far more heavily weighted with theological considerations and warnings of the dire consequences of divine wrath resulting from the sins and vanities of mankind. In the volume's concluding letter to Milner, dated May, 1745, Lobb referred to the recent reappearance of cattle plague and the desirability of reapplying the measures which had been so successful in 1714. He also took the opportunity to state his opinion of the pathogenesis of cattle plague which he compared to that of smallpox and which, Lobb's flowery language apart, is very similar to Lancisi's earlier concept[15].

The immediate source of importation of cattle plague into England in April, 1745, may have been in Flanders; but the disease was by then spreading in many areas in Europe. Early in 1745 it invaded Denmark from the south, and an edict issued in March, 1745, ordered prayers to be said in all churches following its introduction across the border from Germany. Of perhaps more practical value, a temporary ban on markets in the major islands was enforced at the same time. In addition to such governmental precautions, the Danish Academy of Arts and Sciences, then in its fourth year and its second volume of transactions, published three reports on the threatening epizootic, read at the society at meetings in December, 1745, December, 1746, and January, 1746. The two later ones were by professors of medicine who had recently been elected members of the academy, and whose views reflect contemporary thinking elsewhere in Europe on the nature, causation, and possible therapies for the disease[16].

It is the first communication, dated 6th December, 1745, which most accurately records the reactions of the informed layman and cattle owner, in this case one of Denmark's better known writers, the eighteenth century playwright Ludvig Holberg[17]. By 1745 the prosperous author owned two estates, and in other writings from his later years he refers to the devastating losses he suffered to his herds. Holberg stressed the almost 100 per cent mortality in the current outbreak, and the rapid course of the disease when formerly healthy herds succumbed within 24 hours of onset. He also remarked on the specificity of this cattle disease compared to the more general devastation caused by the Black Death in the past. Like other authors before and since, he painted vivid pictures of the plight of many farmers and of the panic caused by failing economy in a country more dependent than most on agriculture and husbandry; although he made a point of pointing out that, in view of the specificity of the disease, its predilection for cattle as opposed to horses was a blessing in disguise. He also, in the mid-eighteenth century, made a point borne out in a later age in cattle raising areas of other continents: on many estates the numbers of heads of cattle had at least doubled over a period of 50 years, with adverse effects on the forests in which the cattle roamed free. The reduction in numbers could give the forests a chance to recover. Like his medical fellow members of the academy in the following year, Holberg came down on the side of contagion as cause of the disease, and supported the theory of a *contagium animatum* or, in the terms of his time, 'insects too small to be seen with the naked eye'[18]. In spite of the sound advice and recommendations offered by Holberg and his two medical colleagues, the waves of cattle plague continued in Denmark and elsewhere on the continent for many years to come.

Similarly, in England, the outbreak which had begun in 1745 continued in a very much more persistent and lethal manner than the previous one, regardless of Lobb's recommendations, and of the letter from Thomas Bates, in *Gentleman's Magazine*[19]. Perhaps its most important and arguably its most interesting result in print came towards the end of the outbreak in Britain, in 1757. The author was Daniel Peter Layard, M.D., F.R.S. (1721–1802), who in 1750 had settled into practice at Huntingdon, where he had rich opportunities to observe 'this calamitous sickness'; in 1757 he recorded his experiences with the disease. In a preface he explained how, when the distemper first broke out in London, he had no access to herds of cattle and did not become involved personally; but in early 1756 there was an outbreak in Godmanchester, near Huntingdon, and Layard vividly described the distress of neighbouring farmers, which compelled him to take an interest in the prevailing cattle disease in an attempt to help. Like many other eighteenth century authors he saw a close analogy between smallpox in man and the present plague of cattle. He even carried his comparisons into the realm of psychology, stating that nursing diseased cattle could more easily be brought to a successful conclusion than could the treatment of smallpox patients, since cattle were 'not subject to intemperance, nor the affrighting passions of the mind', and hence 'nature is neither obstructed nor disturbed by the violent passions of a turbulent or dejected mind'[20].

Layard's main contribution in this treatise was a chapter on inoculation. There is no evidence in his published works, or in the pages of the *Philosophical Transactions*, that he was at any time personally involved in smallpox inoculation; but working in a country practice in the mid-eighteenth century he was certainly conversant with the techniques, the advantages and disadvantages, and the importance of proper preparation of the patient and of judicious choice of time and place for the inoculation. All such considerations he regarded as equally valid for inoculation of cattle, pointing out also that no one should ever consider bringing the disease to areas where it was not already present, and that inoculation should be undertaken only in neighbourhoods where there was an acute risk of infection. All through the ten pages of his chapter on inoculation Layard made copious references to other literature on inoculation of both smallpox and cattle plague. His sources for smallpox are mostly English and French, for cattle plague English and Dutch, culminating in a quote from 'the Dutch professor Swencke' (Thomas Schwencke, 1693–1767) to the effect that 'the beasts do not fall ill till the sixth day, which answers exactly to the observations daily made in the inoculating children'[21].

In 1769 Layard was consulted by the government about a fresh outbreak of the epizootic, reintroduced from Holland to London and hence to Hampshire. It was a limited outbreak and it was soon brought under control by a policy of judicious slaughtering and isolation, as were other subsequent reintroductions in Essex and Sussex from Dutch ports. In April, 1780, Layard summed up his experience of cattle plague in a final letter to the president of the Royal Society, then Joseph Banks at the beginning of his unprecedented more than forty years in that office. Layard was in no doubt of the necessity for tailoring control measures to the different needs represented by different circumstances in different countries. Thus where the disease had become endemic, as in Denmark and in Holland, inoculation of herds at risk held out some hope of control. In England, on the other hand, where only sporadic outbreaks had been experienced for over twenty years, the risk of introducing the disease in healthy areas by inoculation was unacceptably high, and radical slaughter of affected animals was the only solution[22].

In his capacity of authority, after 1757, on all matters relating to cattle plague, Layard also kept in touch with, and was consulted about, the situation abroad. In 1770 he received a royal command to 'hold a foreign correspondence', informing European govern-

ments of the measures which had so successfully controlled
outbreaks in Britain. The worst problems were encountered in the
small countries of Denmark and the Netherlands, on whose
limited territories infection had become endemic. In his letter
Layard described his own rôle in instructing the Danish minister
for foreign affairs, Count Bernstorff, and the court physician, Dr.
Struensee, when they visited England with the mentally unstable
king, Christian VII, in the late 1760s. The resulting inoculation
campaign was said to have been successful, in spite of the
subsequent political fates of the two men: Bernstorff was
overthrown by the intrigues of Struensee in 1770, and Struensee in
turn was executed in 1772, after he overreached himself in
conspiracy against the crown as well as in amorous dalliance with
the unhappy young queen Caroline Mathilde, a sister of George
III. The lessons learned survived, as witnessed by Royal decrees
against cattle plague issued in 1775, 1777, and finally in December,
1778. By October, 1785, they could be suspended, not to be re-
introduced unless '...this distemper should recur, in which case
they will be subject to immediate enforcement'. In the absence of
Bernstorff and of Struensee, the architect of the final plans for
eradication in Denmark was P. C. Abildgaard, since 1773 director
of the Copenhagen Veterinary School[23].

In Holland, early resistance to policies of inoculation and
destruction of affected herds combined with misplaced faith in
spurious 'cures' and 'remedies' to ensure that the disease became
endemic and remained so. The farmers' uncooperative attitudes
foiled all attempts by the inoculators, foremost among whom was
Pieter Camper (1722–89), although Camper received considerable
support for inoculation experiments from a few more enlightened
farmers and other laymen[24]. The versatile Camper, whose artistic
talents made him a master of anatomic illustration, championed a
policy of inoculation against cattle plague both in theory and
practice but was thwarted, according to Layard, by the 'obstinacy
and interruption of the peasants', as well as by the severity and
extent of unfavourable weather conditions. Camper was one more
in the long line of eighteenth century physicians known to
posterity for their skill and perception in dealing with problems of
human disease, who turned with zeal and determination to do
battle with animal pestilence. In 1769, with rinderpest still endemic

in Holland, Camper published a series of lectures on the subject in order to impress on his students their obligation as physicians to watch over not only the health of their fellow citizens, but also equally that of 'all animals useful to the community'. He made no apology for his interest in animal disease[25].

The continual threat of cattle plagues inspired contributions from other European physicians at this time. During the Seven Years War (1756–63) Austria, too, suffered the additional misfortune of outbreaks of rinderpest. In 1762 M. A. Plenciz (1705–86) published in Vienna a volume of observations on infectious diseases in general and smallpox and scarlatina in particular, including also a tract on recent (1755) earthquakes in Europe, Africa and America. Since Roman times earthquakes, and the noxious fumes they released, had been associated with outbreaks of 'contagion'; Seneca referred to 'deadly things... released by Earthquakes to poison the air'[26]. In this tract on contagious diseases Plenciz mentioned the 'dread plague' of cattle which had lasted 'already some thirty years... all around Europe'. By the time his manuscript was ready for the printers, the cattle epizootic had grown to such proportions that he felt the need to add a section devoted exclusively to the *Lues bovinae*. Writing on the eve of the opening of the first veterinary school at Lyons in 1762, Plenciz did not mince words concerning the treatment of infected livestock. He strongly advised physicians and surgeons to use their skills and not leave animal disease to the attentions of 'stupid and untaught veterinarians who know nothing of anatomy or of animal economy, much less of illnesses', and he made a direct appeal for medical men to undertake comparative studies of animals dead of the contagion. Alone among writers in the second half of the eighteenth century Plenciz had no reservations regarding the work of Cogrossi (*cf.* p. 44), believing that small 'worms and grubs' in the ulcerated areas of the mucous membranes of dead and dying cattle might hold a clue to the aetiology of the disease, and that their 'seeds' or eggs could be ingested by healthy cattle at pasture. He speculated on the structure of such eggs, on their tiny size, and on the possibility of organs contained in the foetuses developing from eggs of animalcules of which Leeuwenhoek had asserted there could be at least 2750000 in a single drop of water. Referring to recent inoculation experiments, he concluded that 'like smallpox and measles, cattle plague is infectious and communicable'[27].

In Switzerland in 1773 Albrecht Haller (1708–77), equally famed for his poetry and his botany, and later for experimental physiology, turned to the cattle disease then assuming epizootic proportions in his native country. Reading his account, one might get the impression that he considered the disease to be identical with the cattle plague which had prevailed elsewhere in Europe for the better part of the century; but Fleming in 1871 thought his accurate description of symptoms to be of 'a different disease and in all probability the bovine contagious pleuropneumonia'[28]. Haller's detailed directions for quarantine arrangements, and his theoretical musings on the nature and spread of the disease were sensible enough, whatever the disease in question. He had no faith in 'remedies', and lamented that although the effects of certain simples were known in man, little was known of animal medicine. He wrote: '...few talented persons have observed [animal] diseases; the art of curing them has been left to men of low condition, who have no knowledge of the anatomy of the lower creatures, and who have not informed themselves by the study of nature, or of good authors. The cattle doctors invariably follow the same routine traced by the ancient veterinarians and their science consists of divers receipts, which they have found among the papers of their predecessors'[29]. Switzerland had no veterinary schools of her own until the nineteenth century; the first opened in Berne in 1806, and another followed in Zürich in 1820.

France, on the other hand, had pioneered the institution of veterinary schools, the first opening at Lyons in 1762, the second at Alfort, outside Paris, in 1766. They had been hailed as establishments founded to educate veterinarians capable of solving the problems facing Europe's cattle farmers; they were soon to prove more interested in equine than in bovine species. Nevertheless, the search for answers to the problems of cattle plagues continued; and in the French tradition of prize essays, the Royal Society of Agriculture at Paris in 1765 rewarded Denis Barberet for a tract on epidemic diseases of domestic animals. Although Barberet wrote on several species and their diseases, his main concern was with the prevailing rinderpest; and his common sense advice on control proved so popular that it was reprinted as late as 1808[30]. It is clear from Barberet's writing that sheep pox, which he declared to be identical with smallpox, also presented serious problems in France in the 1760s; and Barberet discussed the differences in specificity exhibited by different infectious diseases, pointing out that although the prevailing rinderpest was specific to cattle, anthrax for example could attack beasts and man with equal facility. He warned of the dangers of failing to destroy carcasses and dying animals, and of using their hides, thus contributing to further spread of the disease. This theme was later taken up by Vicq d'Azyr, who wrote in 1776 of the 'system of killing' that it had by then been 'modified and adapted to the rules of French government'[31].

At the end of the outbreaks in Britain, and towards the end of the century, comments showed that valuable lessons in epidemiology and preventive medicine had been learned, although by the middle of the following century, when rinderpest again reached English ports, they would appear to have been all but forgotten. Certainly in 1771 Robert Dossie (1717–77), like Barberet in 1765, was convinced that rinderpest spread only by 'actual conveyance of the contagious matter', and that consequently isolation or removal of affected animals were the only really effective measures. Although Dossie thought the analogy between cattle plague and smallpox to some extent justified, he saw little value in a policy of inoculation since 'the cause of the difference betwixt the smallpox and murrain be of a dark and inexplicable nature'[32]. Even Linnaeus, preoccupied with developing taxonomy and sparing only a few armchair thoughts for infectious diseases and their causation, noted the effects of cattle plague in Sweden and elsewhere at mid-century. Like many others before and since, he used the scabies analogy in conjecture concerning smaller, invisible beings as agents of disease, although he was content to relegate them to a category he despairingly named 'chaos'[33].

A final comment was contributed by Erasmus Darwin, whose long poem, *The Botanic Garden*, had celebrated Linnaeus' botanical system. In the *Zoonomia*, written between 1794 and 1796, he discussed a number of infectious diseases of animals and man, as well as general points regarding contagion. He included the *pestis vaccina* in a section on bubonic plague in man, and with hindsight he wrote that if the cattle plague 'should ever again gain access to this island', the government must impose strict orders for quarantine and isolation; and that within a five mile radius of any confirmed outbreak all cattle should be 'immediately slaughtered, and consumed within the circumscribed district; and their hides put into lime-water before proper inspectors'[34]. Layard had already set out such precautions in great detail in a chapter 'Of the means to prevent the infection' in his *Essay* of 1757, including suggestions for selection of properly qualified inspectors. Yet, when rinderpest again gained entry into the British Isles in 1865 all this essential advice was all but ignored by the authorities, and by the commercial interests, in spite of renewed more immediate warnings by John Gamgee (*cf*. p. 104).

CHAPTER 5

The first veterinary schools and their corollary: veterinary science in the making

The institution of formal veterinary training in schools established for this specific purpose began in France in the reign of Louis XV. The first veterinary school opened to students at Lyons in February, 1762. The second major French school began giving courses on the Alfort estate at Charenton, near Paris, in October, 1766, briefly preceded by a school at Limoges which, lacking adequate support, lasted only from February, 1766, to November, 1768[1]. Between 1766 and the end of the century, similar institutions were founded in twenty major towns and cities elsewhere in Europe. Some were short-lived, but many survived and are still functioning today. The London Veterinary College (now the Royal Veterinary College) began offering courses in 1791, its first professor being an early, in his native country less than well regarded, French graduate of the Lyons school (*cf.* chapter 6).

The pattern was similar for all the early schools. In most cases governments sent promising young men to Lyons, and later to Alfort, in the hope that they would be able to establish control of animal diseases and to organise similar schools at home on their return. It has been emphasised in many accounts of the beginnings of veterinary education that the establishment of veterinary schools was due to the almost constant fight against cattle plagues in the eighteenth century. There is an element of truth in such statements, but the original French schools owed their existence to a more complex set of circumstances. The man who shaped their development was Claude Bourgelat (1712–79)[2] (Fig. 8). An enthusiastic horseman disenchanted with his chosen profession of

Fig. 8. Claude Bourgelat (1712–79) (courtesy Wellcome Institute Library, London).

the law, he obtained in 1740 the post of director of an academy of equitation, an institution for the training of riding masters, in his native Lyons. Eventually he was also given responsibility for the royal stud farms, and in his extensive work with horses, he became convinced of the need for a new profession of veterinary medicine, with well educated practitioners. Bourgelat began to envisage the creation of a veterinary school, with himself as prospective director and teacher. The idiosyncratic initiative of this ambitious amateur found fertile ground in the intellectual climate of France in the 1750s. It was the decade of the *Encyclopédie*; Bourgelat was a friend of d'Alembert, and contributed articles on all aspects of riding and training of horses, and farriery. Another friend and supporter of his ideas was the physiocrat Henri Bertin who was then Intendant of the Lyonnais. Bourgelat could hope for no better support for his visions of veterinary schools open to serious study of animal disease than that afforded by the French philosophic and economic movements of the eighteenth century.

France also began to acquire agricultural societies from 1757 onwards. Such societies represented a logical extension of the concerns with rural economy of the encyclopaedists and the physiocrats. Their members were less concerned with questions of breeding, and more concerned with present problems of diseases of farm animals, than their English colleagues were to be two decades later (*cf*. chapter 6). There is no evidence of any direct link between the agricultural societies – the Lyons agricultural society dates from 1761[3] – and Bourgelat; but their formation, twenty years before any such developments in Britain, is another indication of the forces at work in mid-eighteenth century France. The appearance of agricultural societies and veterinary schools in France at this time was, in their individual ways, practical expressions of the theoreticians' concern with the importance of a healthy rural economy.

Determined to be suitably qualified to be an architect of veterinary education, Bourgelat enlisted the help of two surgeons, who taught him the rudiments of animal anatomy, physiology, and pathology, to supplement the knowledge displayed in the volumes on hippiatry he had published in 1740 and 1750[4]. In August, 1761, his preparations were rewarded with royal authorisation to open a school 'for the treatment of animal diseases' (Fig. 9). It was established as a private institution, but Bourgelat was granted a subsidy of 50000 *livres* over six years. Classes began in February, 1762, with a total of six students. Numbers of admissions increased rapidly to a peak of more than fifty within a couple of years, only to decline just as rapidly after the opening of the rival school at Alfort in 1766[5].

Ecole Vétérinaire

Fig. 9. The Lyons Veterinary School in the nineteenth century. Engraving in the author's collection.

It soon became clear that Bourgelat's high ideals of a scientifically based training were not easy to attain. Neither he himself nor the few teachers he was able to assemble were particularly well qualified, and the Lyons authorities remained unimpressed and opposed to what was considered Bourgelat's 'pretense to produce scholars rather than practical farriers'. In the circumstances, this was perhaps less than fair judgement, and it did not prevent him developing his ideas further. With the help of his friends in high places he was able to plan another school on the Alfort estate at Charenton, then a few kilometres outside Paris. Bourgelat took over the direction of the new school when it opened in 1766. The Lyons school, smaller, less attractive to students than the Paris establishment, was doomed now to a prolonged struggle with inferior buildings and a much reduced budget until, during the Revolution, a complete reorganisation of the schools began to take place. By then, the organisation of veterinary science and education, and their relationships with the medical profession, had become issues of national concern[6].

Within a few years of the opening of the schools, recruitment of students presented problems to Bourgelat. From the very beginning authorities in major centres elsewhere in Europe were anxious to send students to the French schools in the hope that they would be able to establish similar schools upon their return to their respective countries. As early as September, 1763, the Danish government, inspired by similar Swedish moves and mindful of the continued presence throughout Scandinavia of rinderpest, sent one student of surgery and two medical students to Lyons with a three-year grant. One of the medical students was P. C. Abildgaard, who eight years later became the architect of veterinary education in Denmark, and the first professor of the country's first veterinary school. In letters home from Lyons he expressed his and his fellow students' dismay that they could learn little about practical control of animal plagues. The staff of the school was concerned mainly with 'the theoretical side of veterinary medicine...seeking to explore the discipline suitable for the instruction of the so-called horse doctors, or farriers. The leaders of this school were interested first and foremost in horses, as the most valuable of the domestic animals...'[7]. Abildgaard reluctantly stayed the course, and later received favourable mention as a successful student and even an offer of employment, in spite of his medical background; for Bourgelat had become wary of the presence of medical students in his schools.

In 1772 G. Cicognini, director of the medical faculty at Milan, enquired about the possibility of sending 'young men already well versed in medicine, who might be expected to obtain perfect results' to the Alfort school. This elicited an explosive reply from Bourgelat, who wrote bitterly of past experiences of the 'deranged and debauched' behaviour of well-born young surgeons and physicians, and of his own preference for commoner folk, sons of farriers rather than of farmers, as long as they were able to read and write[8]. In the span of ten years Bourgelat had abandoned some of his early ideas and narrowed the gap between himself and his arch-enemy and rival Etienne Lafosse (1739–1820), who had always advocated instruction in informed farriery rather than Bourgelat's higher visions of a more scientifically based knowledge. Lafosse may have had worries not dissimilar to those of some who today think that students with top marks at school will not necessarily become the best general practitioners or country veterinarians. He wrote: 'Any man who has sufficient intelligence to study veterinary medicine in depth will not consider himself made to be a farrier; he will soon abandon the profession to concentrate on human surgery and medicine...'[9]. In the second half of the twentieth century he will perhaps turn to molecular biology and other research activities.

Neither Bourgelat nor Etienne Lafosse, who, disappointed not to be invited to play a rôle in either of the two schools at Lyons and at Alfort, offered a rival course of lectures of his own, spent much time on species other than the horse. The writings of both men gave prominence to the care of this animal and its diseases. They both wrote extensively on glanders, as had Lafosse's father in a treatise published in 1749, more than ten years before Bourgelat opened the Lyons school to find himself the object of the violent dislike of both Lafosses, father and son[10]. There is ample evidence in these writings of the seriousness of the problems of glanders, especially in mounted regiments in times of war, and also of the difficulties facing contemporary thinking on causation and therapy. Although the younger Lafosse (Fig. 10) – his father died in 1765 – continued his opposition to the established schools, Bourgelat had from the beginning a powerful source of support in Henri Bertin (1719–92). The two men had known each other before the establishment of the veterinary schools, in the 1750s, when Bertin was Intendant at Lyons and Bourgelat director of the riding academy. Bertin was a follower of Quesnay and the physiocratic doctrines of agricultural economics; he took a particular interest in the creation of the veterinary schools, all the more so when in 1763 he obtained the newly created powerful post of Secretary of State for agricultural affairs[11]. His sphere of interests was wider than Bourgelat's, and his policies, aimed at general improvement in agriculture and agricultural economy, helped to convince Bourgelat of the need to be seen to provide some practical aid in rural areas struck by diseases of domestic animals other than the horse.

There were many, frequently severe, epizootics in rural areas in the eighteenth century; sheep and swine suffered as well as cattle, and not long after the opening of the first veterinary schools their students were called upon to show, rather prematurely, the results of the instruction they were receiving. The Lyons school had opened to students in February, 1762. In July of the same year an epizootic, affecting horses and cattle, struck in the Dauphiné region. Bourgelat, never inclined to hide his light under a bushel, and mindful that his government support originated in hopes of eventual control of epizootics, immediately set forth to inspect the stricken area. He took with him a group of seven students, many young teenagers, who were left there on their own with instructions to attempt to control the outbreak. The 'chef de mission' was Louis Bredin (1738–1813), who was later to become director of the Lyons school (1780–1813) and steer it through its financial and administrative difficulties both before and during the Revolution[12].

The first mission was spectacularly, if probably fortuitously, successful in saving a majority of the remaining affected animals through intensive nursing, and probably because the group was fortunate to arrive at a time when the outbreak was in any case on the wane. The success was a useful boost for the young school, and helped to earn for it ultimate approval when Louis XV granted it a Royal Charter in 1764. Later the early promise could not be sustained and Bourgelat, his students and successors both at Lyons and Alfort found themselves powerless in the face of the severity

Fig. 10. Phillipe Ètienne Lafosse (1739–1820); engraving by
Michel after Harquinier (courtesy Wellcome Institute Library,
London).

of later outbreaks of cattle plague. During the 1770s it became increasingly obvious that their attempts to control a particularly persistent and extensive epizootic of rinderpest were lamentably ineffective. At the same time politics, and medical politics, were working against them. When Bourgelat died in 1779, his old ally Bertin, the committed physiocrat whose administration had initially provided the funding for the veterinary schools, was no longer in a position of power. Later incumbents of his office could not, or would not, support the veterinary schools to the same extent; reforms in veterinary courses introduced by Bertin before he finally left the ministry in 1780 did not survive the next administration. There was no lack of medical opinion seeking to persuade the politicians that physicians were more competent when dealing with epizootics as well as epidemics, and that medical and veterinary teaching should come together, albeit with the emphasis on medical and surgical practitioners and teachers. In the last decade before the Revolution, and in its early years, powerful forces were at work to attempt an integration of medical and veterinary teaching in a higher unity which should be economically and philosophically justified. Talleyrand declared in September, 1791, that it should be self-evident that veterinary medicine and surgery must necessarily be united with human medicine, since the great basic principles of the Art of Healing never change, only vary in their application[13]. Barely come of age, the veterinary schools had a fight for survival on their hands.

In the year 1774, France had faced problems of disease in both animals and men. In addition to the prevailing rinderpest, there was a serious outbreak of smallpox. Traditionally the French had been less willing than most of their European neighbours to subject themselves to smallpox inoculation. The king was no exception, and in May, 1774, Louis XV caught the disease and died, causing the rest of the royal family, and many of their subjects, to revise their attitude to inoculation[14]. Rinderpest also raged unabated when in the same year A.-R.-J. Turgot (1727–81) took over Bertin's former post as Comptroller-General of Finances. Like Bertin, Turgot was a physiocrat with a passionate belief in the importance of improvements in all areas of agriculture, including the veterinary schools. Also like Bertin, his idealistic belief in improvement, regardless of the government's immediate financial difficulties, brought about his fall from power, in his case after less than two years in office, in 1776. It was Turgot who, ten years earlier, had founded the short-lived third veterinary school at Limoges, when he was Intendant of that region. Dependent on local funding it was a failure and closed after less than three years. During his period in charge of finances in Paris he renewed his efforts to help rural areas in their fight against the current rinderpest, and against epidemics and epizootics in general.

With this purpose in mind Turgot appealed in 1775 to the *Académie des Sciences* for assistance, and requested that it send two members to study, and attempt to control, the cattle plague which was rampant in several provinces in the south of France. The academy sent the recently elected Félix Vicq d'Azyr (1748–94); having invoked traditional measures, he could report the outbreak over in a matter of months. From then on, in a short life, he was to have a lasting effect on official attitudes to infectious diseases of animals and man in France and elsewhere. On his return to Paris from his successful mission to the stricken countryside he found a receptive administrator in Turgot, and in April, 1776, a Royal Commission was appointed to deal with the problems of epidemics and epizootics in France. Turgot's ministry did not long survive this event; other reform measures of his were highly unpopular, and he was removed from office a month later. But the commission, under the able leadership of Vicq d'Azyr, continued to flourish, and in 1778, as the *Société Royale de Médecine*, it received its first letters patent[15].

Within the framework of this society Vicq d'Azyr created a new approach to public health and hygiene and laid the foundations for the comparative medicine of infectious diseases which was to develop so successfully in the following century. The seeds are there to be seen in an early memorandum from the society, which in 1780 distressed Philibert Chabert (1737–1814), who had succeeded Bourgelat as director of Alfort in spite of his lack of obvious academic qualifications. He was a farrier who had taught shoeing competently enough at Alfort from 1766; in 1783 he wrote a treatise on anthrax in animals (*cf.* chapter 7, p. 125). Chabert objected to the society's desire to involve medically trained surgeons in the teaching in the veterinary schools, fearing that this would only serve to turn students in the direction of human medicine. There is a prophetic sentence in the memorandum which shows that Vicq d'Azyr and his fellow officers in the society were well aware of the direction in which they were moving, although their aims were to be fully developed only in the following century, after the Revolution and the Napoleonic wars. They wrote: 'Just as there is in existence a comparative anatomy, it should be possible to establish also a *comparative medicine*...[16].

Until illness, coinciding with the Revolution and the Terror, put a brake on his activities, Vicq d'Azyr wrote extensively on the recurrent outbreaks of cattle plague; he was also either personally involved, or gave encouragement to other members of the society who were consulted concerning outbreaks of disease in animals and man. The subjects covered during the society's existence, from 1778 to 1792 when the revolutionary regime suppressed all corporations and societies, spanned a wide range of disease outbreaks, from cattle plague and sheep disease to dysentery, 'miliary fever', ergotism, and intermittent fevers. The report on the latter, signed by Vicq d'Azyr and Jeanroi, referred to the age-old problems regarding the draining of marshes, which had for centuries been suspected of harbouring the cause of the disease. It also instanced a number of localities in Europe, America and Africa in support of the 'well-known fact' that a combination of high humidity and warm air increases the severity of the problems of intermittent fevers[17].

In 1782, after several abortive attempts, Vicq d'Azyr was able to convince a sympathetic administration of the need to develop a more scientific outlook within the veterinary school at Alfort. The school acquired three new professorial chairs: one in comparative anatomy for Vicq d'Azyr, one in rural economy, and one in chemistry, and in addition a complement of new research facilities. This resulted in achievements which helped to make veterinary medicine into an acceptable subject for research; unfortunately, there was no comparable improvement in the teaching. A contemporary critique of the situation came from Arthur Young, founder-editor of *Annals of Agriculture* (*cf.* chapter 6), who visited the Alfort school in 1787. Young admired the school, and the farm attached to it; but he had reservations about the combination of high-powered academics and agricultural studies to be found there. His tactfully oblique criticism of the running of the farm by

Daubenton, 'high in royal academies of science' and lecturing on rural economy, suggested that it might be such men, 'celebrated through Europe for merit in superior branches of knowledge' could hardly be expected to turn out 'good ploughmen, turnip hoers, and shepherds'[18]. The project was indeed discontinued in the same year; the school lost its research professors and reverted to its old ways. This misfortune did not deter Vicq d'Azyr, who believed firmly in the unity of human and animal medicine, and who had also observed that through veterinary research might be gained the opportunity to carry out certain 'valuable and bold experiments which would be criminal if attempted in the treatment of human disease'[19]: in other words, the use of animal models.

With Vicq d'Azyr and the *Société Royale de Médecine* on the one hand, and with the developing veterinary schools on the other, there were in pre-revolutionary France two trends evolving side by side, and running parallel for many years. Young veterinary practitioners in rural areas were often inclined, and sometimes encouraged, to offer their services to local peasants as well as to their animals, when there were no resident physicians in the area. Many saw it as a move towards medical rather than veterinary practice, and upward social mobility[20]. In contrast physicians involved in animal disease research, who had no reason to worry about their social mobility, were simply keen on the scientific aspects and interested in comparative research. Increasingly, as veterinary schools became established throughout Europe, such physicians, in France and elsewhere, came to regard additional veterinary knowledge, and sometimes even qualifications, as a means of broadening their outlook and understanding.

Vicq d'Azyr died in 1794, in the penultimate month of Robespierre's reign of terror. His early death has been accepted as due to natural causes, possibly tubercular in origin. By conviction, and as erstwhile physician to Marie-Antoinette, his attitude to the Revolution was ambivalent. It has recently been suggested that, dismayed by the turn of events, he may have followed the example of his friend Condorcet (1743–94), who had escaped the guillotine by taking his own life, by poison, two months earlier[21]. Still in its infancy, comparative medicine suffered a setback with his death and with the revolutionary dissolution of the society he had led through such promising beginnings; but the society was to be

Pl. 1.

Fig. 11. The dual functions of shoeing and surgery in a French *maréchalerie* as depicted in Diderot and d'Alembert's *Encyclopédie* (courtesy Wellcome Institute Library, London).

resurrected after the Revolution in the form of two institutions of
lasting importance, the *Académie de Médecine* and the *Société de
Médecine de Paris*. The seeds of animal experimentation and of
comparative medicine on a scientific basis had been sown and were
to develop in the next century; there were other pointers as the
eighteenth century was drawing to its close.

In England, veterinary education developed relatively slowly
(see chapter 6). After initial steps had been taken by an agricultural
society, the initiative passed to political factions, and courses
began only in 1791, given by the Frenchman Charles Vial de
Sainbel. He was a graduate of the Lyons school, whose career in
France had been beset with difficulties; frowned upon by Bourgelat
and his successors for an unfortunate mixture of minimal
competence and unwarranted arrogance, he had been unable to
hold down any appointment for any length of time, and had come
to London to seek his fortune[22]. When the college opened its doors
to students, Sainbel was its one and only professor; and when he
died suddenly in August, 1793, probably of glanders, veterinary
education in London might have been suspended indefinitely.
Temporary rescue was at hand, in the person of John Hunter
(1728–93).

Hunter, like Vicq d'Azyr, had a lasting and deep-seated interest
in comparative anatomy; he acquired dead animals from the
Tower of London Zoo for his private museum, and he was
reported to sometimes see sick animals in consultation. In 1790 he
had encouraged the young William Moorcroft (1765–1825) to add
to his medical qualifications by a period of study at the veterinary
school at Lyons. In March, 1791, he was elected an honorary
member of the London college, which he from then on supported
in every way. At the sudden death of Sainbel he opened his own
lectures to veterinary students, as did his friends and colleagues
Matthew Baillie (1761–1823) and George Fordyce (1736–1802).
This move may have ensured a better quality of education for that
particular group of students than they would otherwise have had,
although Hunter's own death barely two months later again left a
gap[23].

John Hunter was best known as surgeon and anatomist, but his
range of interests was wide, and included transmissible diseases.
His early encouragement of Edward Jenner (1749–1823) is well

known; had Hunter lived, he might have advised a more thorough experimental basis for Jenner's first publication on the *variolae vaccinae*, and some of the early uncertainties and criticism could have been avoided. Instead, appearing as it did five years after Hunter's death, Jenner's *Inquiry* was an inspired piece of empiricism which will remain a pillar of immunology as well as the basis for the tool which eventually, more than 150 years after Jenner's death, made the eradication of smallpox possible[24]. Jenner's strength was his practical application and tireless promotion of the new method, not the depth of his initial documentation[25].

Ten years after his introduction of vaccination, Jenner briefly turned his attention to a disease which had for some time presented problems to keepers of hunting and shooting dogs in France and in England, and to human patients bitten by infected dogs, because of disturbing confusions with symptoms of rabies. Dog distemper was, Jenner wrote, 'as contagious among dogs, as the small-pox, measles, or scarlet fever among the human species'. He referred to his own vain attempts to eliminate the contagion by white-washing and fumigation of contaminated kennels, a method which in the previous century had been used successfully in France[26]. He recorded the lasting immunity of those dogs fortunate enough to recover, and commented on the specific susceptibility of dogs to the disease, and its non-transmissibility to man.

In 1783 John Hunter had founded, with George Fordyce, a small exclusive London society. With a maximum membership of twelve, it was more of a club where Hunter and Fordyce met with like-minded friends and colleagues, many of them Hunter's former pupils, to discuss cases and other matters of professional interest. The first volume of transactions appeared ten years after the initiation of the society, in the year of Hunter's death. Of the eighteen papers included, four were by Hunter himself; the penultimate paper of the volume appeared under the name of John Hunter, M.D. (1754–1809), Hunter's younger namesake[27]. Judging by the title of the paper, and its constant emphasis on cooperation by all members of the society, it would seem to have been a joint effort, finally written up by John Hunter, M.D. It may or may not have been the idea of the elder John Hunter, who is known to have counted rabies among his interests[28]. Whatever its origin, the resulting document not only sums up the state of knowledge of rabies towards the end of the eighteenth century; it gives explicit suggestions for animal experiments which might serve to improve understanding of the disease. Referring in standard eighteenth century fashion to the 'saliva, or poison' as the transmissible factor, Hunter advocated inoculation of healthy dogs, and also other animal species, with saliva from dogs known to be rabid, to enable observation of the disease under controlled conditions. Such experiments would also afford the opportunity of studying the effects of removal of the inoculation site at various intervals after the introduction of saliva, and of the application of what the author called 'counter poisons'. The final suggestion was for the inoculation of a healthy dog with saliva from 'an hydrophobic patient'[29].

None of the members of the society took up the challenge. John Hunter suffered a fatal heart attack in the same year, and John Hunter, M.D. published nothing further of importance between 1793 and his death in 1809. The practical execution of the experiments suggested in London in 1793 was to come in the following century, in Germany and in France. France above all was to be the country most closely associated with the development of animal experimentation in the nineteenth century; and in France there had been a forerunner to Hunter's theoretical suggestions, albeit concerned with a different contagious disease.

In 1765, with his Lyons veterinary school firmly established and plans for the second one at Alfort well advanced, Bourgelat published at Lyons a *Materia medica* for the use of his students. It was an early veterinary pharmacology text *cum* pharmacopoeia, with medicinal prescriptions as well as information on the effects, according to Bourgelat, of the various drugs presented for the use of his pupils. The book contains a good deal of extraneous advice and anecdotal name-dropping, which would appear to be included as proof of the author's eminent credentials and important connections. It is in this latter category, in a section on glanders, that he included 'for the instruction of the pupils', certain ideas for transmission experiments with the disease. They were prompted by a prophylactic remedy, of undisclosed composition, developed by a certain Baron Sind, a cavalry officer in the service of the Elector of Cologne. Sind's remedy and its distribution followed a similar pattern to the many 'cures' for human ills offered in the eighteenth century[30]. Bourgelat appears to have had no illusions concerning the efficacy claimed for this 'remedy', but he responded to the enquiries of a French diplomat accredited to the court at Bonn, who proposed to test the prophylactic in experiments on nineteen horses. Bourgelat explained the recommended experiments in careful detail. He emphasised the necessity of ensuring that the horses selected as suitable for infecting others were genuinely glandered. This condition cannot have been easy to fulfil at the time, with a disease of notorious diagnostic difficulty, although it was distressingly common in the horses of mounted regiments. He proposed attempting to infect eight of the nineteen animals available after they had been given the prophylactic remedy, exposing eight others without any protection as controls. The remaining three could be kept as healthy controls, to be introduced later into contaminated stables[31].

Bourgelat made no reference to any results of the suggested experiments; but Erik Viborg (1759–1822), Abildgaard's successor at the Copenhagen veterinary school, later examined in some detail a French report of such a test of Sind's remedy. Although Sind himself claimed to have demonstrated the efficacy of his product in a series of experiments such as suggested by Bourgelat, Viborg was more than a little sceptical. He detected several inconsistencies in the report, and a certain bias on the part of the farriers, all employed by Sind, who had carried out the experiments. In fact, Viborg accused Sind and his henchmen of deviousness and deceit with regard to the declared results. He named a number of authors, including Abildgaard, who had been entirely unable to confirm the putative protective effect of Sind's remedy[32].

Viborg had himself carried out a few transmission experiments with glanders in the 1790s, with positive results. Inoculations were apparently made whenever a suitable animal was conveniently available: that is, a fairly old, more or less emaciated, but otherwise sound, horse in the stables of the school, which would be not a great loss when sacrificed[33].

The members of Hunter's Society for the Improvement of Medical and Chirurgical Knowledge had regarded rabies as a generally transmissible disease, but had not entirely ruled out spontaneous occurrence in dogs. A contemporary who had no doubts at all was Samuel Argent Bardsley (1764–1851) who, also in 1793, published his observations on rabies and hydrophobia in an early volume of the *Memoirs* of the Literary and Philosophical Society of Manchester. Bardsley spent his working life as physician to the Manchester Infirmary from 1790 to 1823. It was a time when the threat of urban rabies was never far away, and Bardsley had no lack of case notes on which to draw for his firm views on the disease and its prevention. He believed that canine madness, and hydrophobia in man, was a transmissible disease which could not occur spontaneously under any circumstances. He wrote: 'It is confidently asserted, there never existed in any case, the marked and decided characteristic symptoms of genuine canine madness, without the intervention of the poison of the rabid animal...'; and later: 'That a specific virus, resembling the canine, is spontaneously generated into the human system, – Such an Hypothesis is alike unsupported by fact or analogy'[34]. He was never to alter his stand. Seven years into the new century he published a volume of case histories and other observations culled from his hospital practice. It included his earlier paper on rabies reprinted in its entirety, with added thoughts on a 'plan for its extirpation from the British Isles'. Having established what he considered 'an accumulated series of probable evidence' for the impossibility of spontaneous occurrence of either canine madness or hydrophobia in the human subject, he presented a plan:

> ...as simple as I trust it will prove efficaceous. – It consists merely in establishing an universal quarantine for dogs within the kingdom, and a total prohibition of the importation of these animals during the existence of such quarantine.

He compared the possible benefits of such rules with the potential of Jenner's recently introduced vaccination, referring to rabies as 'so tremendous a pest to society'; all of which serves to emphasise the extent to which the disease was still perceived as a major threat. He continued:

> Our insular situation is peculiarly favourable for the experiment; and the present period most propitious for the attempt. The alarm from the extensive spread of canine madness in the south of England is at its height;...[35]

It was to be another 80 years before Pasteur voiced similar views, and caused implementation of such legislation in England, resulting in the eventual eradication of the disease in the British Isles (*cf.* chapter 10). It should perhaps be noted that Bardsley concluded his 'thoughts on canine madness', in 1807, with an appeal for comments on his proposals from 'any medical gentleman who has perused the foregoing remarks, and directed his attention to the general subject'. It may be significant, in view of the state of the veterinary profession in England at the time, that there is no similar appeal to 'veterinary gentlemen'.

In the last decade of the eighteenth century, when rinderpest had seemed to be on the wane for some time, there was a final serious outbreak in Italy, where the European epizootic had begun in 1711. The centre of the outbreak this time was in Piedmont. It was examined in detail in two tracts, published in 1793 and in 1798, by Michele Francesco Buniva (1761–1834) (Fig. 12). Buniva was a graduate of Turin's medical college, who remained there to teach after completing his own studies in 1788. His abiding interest in comparative studies led him to develop close links with the veterinary school that had existed in Turin since 1769. Its director was C. G. Brugnone (1741–1818), who had been among the first foreign (medical) students at Lyons and at Alfort, and who had taught in the school since his return from France. The working relationship of the two men, within the same academic environment, formed a basis for the development of comparative medicine there[36].

Buniva's writings on rinderpest were mainly concerned with questions of control, and included detailed descriptions and examples of the many possible pathways of transmission, among which he noted the sexual one: bulls infecting cows during copulation. He rejected at some length the idea of aerial transmission, in particular the theories of Samuel Latham Mitchill (1764–1831). Mitchill had claimed to have proved that the 'miasma of contagion' derived from Priestley's 'nitrous dephlogisticated air', and that it was especially prevalent in the atmosphere when outbreaks of contagious diseases occurred. His was a variant of Sydenham's *genus epidemicus*, adapted to the new chemistry of the late eighteenth century[37].

The shadow of the French Revolution had hung over Vicq d'Azyr's last years; its reverberations caught Buniva at the peak of

MICHEL   BUNIVA

Fig. 12. M. F. Buniva (1761–1834) (courtesy Wellcome Institute
Library, London).

his career. Innocent of politics, and with connections in France
among both medical and veterinary fraternities, Buniva was forced
into temporary exile in France in 1799, when his French sympathies
proved inimical to the prevailing political climate in Turin. Until
he was able to return, shortly after the turn of the century, he spent
his time well, absorbing the latest veterinary knowledge at Lyons
and Alfort. Nor did he neglect medical concerns. He took the
opportunity to see for himself, travelling in France and in England,
the effects of Jenner's new vaccine; when he returned to Turin he
carried with him the knowledge and the material to begin
vaccination on his native soil. His contributions were both
theoretical and practical; he organised a free vaccination service in

Turin, which continued to benefit from his skill and experience throughout his lifetime[38].

While he was still in Paris, in 1800, Buniva read a paper to the *Société de Médecine*, now functioning again, without its royal prefix. From the title of the paper, and its first paragraph, it is clear that the Society had lost none of the zeal for comparative medicine which it had enjoyed under Vicq d'Azyr. The subject was an extensive epizootic, probably of feline distemper[39], which had for some years been decimating cat populations in France, Germany, Italy and England. Buniva wrote: 'Epizootics in general, and the present one in particular, demand the attention of the medical profession; they are of direct relevance to the human species'. He pointed to the grave results for towns and countryside alike when the demise of the majority of cats resulted in an upsurge in the number of rats; he presented results of extensive *post mortem* pathology; and he was at pains to reassure the society, and the public, that the cat disease was not identical with the recent bovine plague. They were separate diseases, specific to the respective species, and not transmissible to other species, and certainly not to man. Although he had no time for Mitchill's miasmatic theories, and talked freely of the 'contagious principle', Buniva could not see eye to eye with those 'partisans of the *pathologia animata* who believed that the cat epizootic, and other contagions of animals and man, were caused by 'infinitely small animalcules'. He claimed to have made experiments and observations which ruled out such an explanation[40].

Michele Buniva's life and career spanned in almost equal parts the later eighteenth century and the pre-Victorian years of the nineteenth. In his breadth of vision, his versatility, and his veterinary sympathies, he was representative of a new generation of medical and veterinary scientists, which was to transform the study of infectious diseases for the remainder of the new century.

CHAPTER 6

Patterns of veterinary education and professional achievement in England 1750–1900

The rapid adoption and development of Bourgelat's ideas in France and in other countries on the European continent did not take place in Britain, although the opening of the Lyons and Alfort schools was certainly noted in London. A few years earlier, in 1758, a group of London farriers led by one John Wood had publicly suggested the establishment of a hospital for horses in view of the generally 'low state of the art', but had found no support for their ideas. More importantly, following the opening of the French schools, Edward Snape, farrier to George III, made a comprehensive proposal in 1765. He outlined plans for an 'Hippiatric Infirmary' with a school attached with facilities for 'instruction of pupils in the profession'. He also presented plans for lectures on scientific subjects as well as clinical instruction. It took Snape twelve years to finalise his plans and to solicit enough promise of support; but when the school at last opened in 1778 the promised financial support, by subscription, fell away, and the scheme did not survive[1]. This is all the more surprising as George III in the same year signed an authorisation for the establishment of a veterinary school, the first in the German states, for Hanover, whose absentee Elector he remained[2]. The short-lived English plans were still concerned only with horses. Cattle and other species were not mentioned.

By the time Snape's infirmary plans collapsed, there were already more than a dozen new schools and university departments offering veterinary education all over Europe. The reasons for the tardiness of official response in England were complex and not immediately obvious. Pugh thought to find one reason for the almost thirty years' delay in the creation of a veterinary school for England in a 'contrast between the benevolent despotism of the Continent and the *laissez-faire* regime of eighteenth century England', with the former including animal medicine in a general scheme of economic reform, and the latter requiring private initiative and funding for any reform. It may have been less straightforward. When Snape's subscribers failed to meet their obligations in 1778, there was already a lively interest in general reform in agricultural matters in England. Arthur Young (1741–1820) had begun his writings on the state of agriculture and agricultural reform at an early age in 1767[3], and was later to play his part in the introduction of veterinary education in London. Throughout the country the old common land system was gradually giving way to a different kind of arrangement with individual holdings of separate fields; fields which were beginning to show the benefit of new methods of crop rotation, proper drainage, and improved cultivation in general[4].

The growing interest in improvements to agriculture was also manifest in the formation of agricultural societies, although they, too, began their existence nearly two decades later than their French equivalents. The Bath and West of England Society, which still exists, was founded in 1777, and others were soon to follow. But in their early years these societies placed more emphasis on improvements in yields, and in the quality of stock and of new breeds of domestic animals, than on any attempts to prevent or treat their diseases. It might have been thought that the king nicknamed 'Farmer George' would have been as enthusiastic a supporter of his farrier's plan to open an educational horse infirmary as Louis XV and his ministers had been for Bourgelat's schools; but the 1770s and 1780s were troubled years for George III, and his signature on the Hanover document may have been only in the nature of a rubber stamp. The king's mental illness first surfaced in 1765; the war with the American colonies accentuated his problems at home. When he did indulge his taste for farming, his interests, like those of most gentlemen farmers at that time, focused on new breeds and better crops[5]. The big landowners and their tenants, meeting in the agricultural societies, reserved their enthusiasm for breeding improvements and ploughing matches, spinning competitions and cattle shows. The medical profession had coped ably with the problems of cattle plagues which in any case, after 1770, were no longer acute. Farriers and cow leeches, however incompetent, were left to treat such problems as remained as best they could.

During the later 1780s the ideas emanating from the French veterinary schools were increasingly being disseminated; and in the wake of the failure of Snape's project they finally, albeit slowly, began to gain ground even in England. The first professional body to seriously turn its attention to the need for formal veterinary education in Britain was the Odiham Agricultural Society of Hampshire. Six years younger than the Bath and West of England Society, it was founded in 1783; unlike its senior, it was not to survive into the nineteenth century. But in the relatively few years of its existence it achieved what none of the other societies had even attempted: the founding of a veterinary college for London, the first in the country. The Odiham Society was fortunate in its early membership. A founder member was Thomas Burgess, the youngest of three sons of a local grocer. Educated through scholarships to Winchester and to Oxford, he was eventually to end a distinguished career as bishop of Salisbury. He may have joined the society to further an interest in the moral improvement of farm workers; once there, he turned his concern to the animals of the farm, and their care in sickness and in health. It was Burgess who first proposed that the society should consider the possibilities of improvement to farriery through the medium of formal education, and that such education should be based on '...study of the anatomy, diseases and cure of cattle particularly horses, cows, and sheep,...'. This, said Burgess's motion, would benefit both agriculture and other 'important branches of national commerce'. A humane theologian, Burgess was not blind to economic arguments.

Burgess's proposal was dated 19 August, 1785; the road ahead was to be long and arduous, and the initiative was gradually to pass from the society to other hands. Meanwhile its members agreed to establish a 'Farriery Fund' of voluntary subscriptions in order to further breeding, management and general improvement of horses, cows, sheep and swine, and to search for cures for diseases. For the first time, concern with the health of farm animals was publicly expressed. Adequate management and registration of dairies and of profits and losses in herds and flocks were also among the aims of the Farriery Fund when it was launched in June, 1786. There the matter rested until the initiative of Thomas Burgess was reinforced by the report of a visit to the Paris veterinary school of Alfort made by Arthur Young in October, 1787.

Since Bourgelat's death in 1779, the director of the Paris school was Philibert Chabert who personally conducted Young around the school, its classrooms, its laboratories, and its farm. The latter had been acquired only four years earlier, in 1783, at the time when the school had also added four new teaching posts: two in rural economy, one in anatomy, and one in chemistry. Young was more impressed by the school and its teaching than by the management of the farm. He wrote approvingly of the dissecting rooms and of the large collection of preserved samples of 'the most interesting parts of all domestic animals', and of pathologic specimens showing 'the visible effect of distempers'. All this, he was assured, for the moderate sum of £2600. On the other hand Arthur Young, country squire and writer on agricultural reform, had very mixed feelings about the state of the farm, presided over by the academically able Louis Daubenton (1716–99), Buffon's erstwhile assistant. Young's report reflected the informed outsider's view of the results of Vicq d'Azyr's attempt to introduce a more intellectual approach to teaching in the French veterinary schools. He wrote:

> There are at present about one hundred *élèves* from different parts of the kingdom, as well as from every country in Europe, *except England*; a strange exception, considering how grossly ignorant our farriers are; and that the whole expense of supporting a young man here does not exceed forty *louis* a year; nor more than four years necessary for his complete instruction. As to the farm, it is under the conduct of the great naturalist [Daubenton], high in royal academies of science, and whose name is celebrated through Europe for merit in superior branches of knowledge. It would argue in me a want of judgment in human nature, to expect good practice from such men. They would probably think it beneath their pursuits and situation in life to be good ploughmen, turnip hoers, and shepherds; I should therefore betray my own ignorance of life, if I was to express any surprise at finding this farm in a situation that I had rather forget than describe. ...[6]

Arthur Young's on the whole favourable impression of the French system was widely reported on his return home, and was not ignored by the Odiham Society. In May, 1788, it was decided to send two students to Alfort. It was a beginning, of sorts; there was no mention of the students eventually being able to teach others at home, as had been the case elsewhere in Europe. It was to be another two years before the already medically qualified William Moorcroft (1767–1825) went, on the recommendation of John Hunter, to study at the Lyons veterinary school. Also in 1788, James Clark, Scottish farrier to George III, commented:

> ... a regular mode of education is not to be attained on any terms, at least in this country. In France, a regular academy for the instruction of young Farriers has been instituted. The attempt is laudable, and worthy of imitation. The Physician and Surgeon enjoy the greatest opportunities for receiving instruction in their professions, by regular education. The analogy between the diseases of the human body and those to which the horse is liable, is very great. Hence it must be obvious that the cure of those diseases which affect the latter, must depend upon the same principles as that of the former: from which it is likewise evident that a regular education is necessary to a Farrier.[7]

All these well meaning suggestions were in need of a more determined driving force behind them. It came in the person of Granville Penn (1761–1844). Unlike his father and his grandfather, the founder of Pennsylvania, Granville Penn preferred the life of a Greek scholar, albeit with an admixture of the eighteenth century man-about-town, to the more adventurous existence of the colonist. He was, however, a natural campaigner, ready to take on worthy causes; and, financially independent, he was in a position to do so. Through contacts in Oxford and in the Society of Arts, Penn developed an interest in the Odiham Society and its efforts in the promotion of veterinary education in England. He had become a member and a subscriber to the Farriery Fund when in August, 1789, the society passed a resolution stating that 'It is to be lamented that there is not yet in England any Establishment adequate to the desired improvement of Farriery by a regular education in that Science'[8].

At this point fate played into the hands of Penn and of the cause of veterinary education in London. In his perambulations around London society Penn was, by chance, introduced to a certain Frenchman, one Charles Benoît Vial de St Bel. He had not always had so imposing a name; at home in France he had been known

– not always favourably, it must be said – as plain Benoît Vial, from the village of St Bel. To posterity he was eventually to be Sainbel. He had been educated at Lyons and Alfort; the college register of alumni at Lyons later described him and his change of name as 'false and ostentatious'; the director at Alfort, Bourgelat, described his character as 'scheming and detestable'. Later he tried, and failed, to establish a veterinary course at the University of Montpellier, where the authorities deemed him 'unstable and not entirely reliable'. Finally, after his death in London, Bredin, director at Lyons, had the last word: 'He began in Lyons, in Paris, in Montpellier and in England, a number of projects, and in all of them he failed. Full of conceit and arrogance. His funerary honours went far beyond his deserts.'[9]

Sainbel's first visit to England in June, 1788, was unsuccessful professionally. He could find no support for the plans he had conceived for an English veterinary school; but in the course of his visit he acquired an English wife, and although they went to Paris at the end of the year, they were soon back in her native country. His claims then to be a refugee fleeing the French Revolution were probably without foundation; but the death of the famous thoroughbred 'Eclipse' in O'Kelly's stud[10], which Sainbel had included in his previous visit, changed his fortune. He was asked to dissect the horse, and his report on the dissection brought his name to the attention of owners of other thoroughbreds. He still experienced some difficulty in finding support for his schemes, until he met Granville Penn. For once, he found a sympathetic listener. Penn realised that here was a man who could head the kind of organisation he had in mind. Explaining his motives tactfully to the touchy and excitable Frenchman, he rewrote and fashioned Sainbel's 'Plan' for a school to suit an English readership. It became an admirable blueprint for a veterinary school, and for the profession which it was confidently expected would result, even to anticipating formal incorporation by Royal Charter. For a number of reasons, there was to be a long wait for that desirable outcome[11].

The revised plan was Penn's manifesto and campaign document. Individual copies were distributed to leading livestock owners, and to the various agricultural societies now flourishing up and down the country. The Odiham Society responded by passing a resolution that '...such an Institution for Education in Farriery is necessary, as has been established in France, Germany, Piedmont, Sweden, Denmark, etc.'. The Society also established a London Committee, thus acknowledging Penn's assertion that the ideal place for a school would be in London. In accordance with contemporary custom, the deliberations of the committee took place in various London coffee houses late in 1790[12]. After collection of the necessary support in word and in deed, and in promises of subscription, the college was ready for its first pupils early in the following year (Fig. 13). The project soon acquired a number of patrons among wealthy and aristocratic landowners; the committee changed its name to 'The Veterinary College, London', and the Odiham Society gave its blessing to the college as an independent establishment[13].

Throughout 1791 support for the new college grew. The presidency was accepted by the Duke of Northumberland, and medical and scientific backing was ensured by the election as 'honorary and corresponding members' of Sir George Baker, John Hunter, and Sir Joseph Banks[14]. In January 1792, Sainbel began lecturing to four resident pupils; by the end of the year the number had risen to fourteen. True to his French training Sainbel was adamant that medical students should be excluded from his lectures, although he had no objection to teaching his own students human surgery. In London as in France it was feared that the influence of medical students would be disruptive and would deflect budding veterinarians in the direction of human medicine as a more 'honourable' profession. The first year in the life of the college was beset by other problems. Some were financial; a programme of building to improve accommodation for the school, its teaching, and its animal hospital soon exceeded existing resources. Then it was discovered that certain pupils had left without completing their course, but were nevertheless claiming to have graduated as qualified veterinarians[15].

In August, 1793, all other problems were temporarily eclipsed when Sainbel died after a short acute illness, perhaps glanders contracted from infected horses in the stables of the college[16]. His tenure had not been unblemished. His competence had been questioned (his poor command of English added to the students' difficulties), but an investigation had largely exonerated him. The recommendation that his teaching load should be shared had not been followed up, perhaps chiefly for want of suitable candidates

Perspective View of the VETERINARY COLLEGE.

Fig. 13. Early view of the London Veterinary College (courtesy Wellcome Institute Library, London).

for a second post. Now the future of the school was more than ever in jeopardy. Even the first intake of students had far from completed their three year course. An intensive search for a successor to Sainbel began immediately, conducted by a group of subscribers to the school which included Sir George Baker and John Hunter, and others with a commitment to comparative studies. John Hunter also made a contribution to bridging the gap for the students by admitting them to his own lectures, and prevailing on other medical teachers to do the same[17].

When Hunter himself died less than two months later, the crisis deepened. The need for a speedy solution produced a plan for the administration of the school and its teaching to be shared by two men, William Moorcroft and Edward Coleman. Moorcroft, a surgeon with added veterinary qualifications obtained at Lyons on the recommendation of John Hunter, was at the time practising in Oxford Street as the country's only qualified veterinary surgeon[18], and showed only reluctant interest in a possible post at the school. Coleman, who keenly pursued the appointment, was also trained in human surgery but had no veterinary qualifications of any kind. The two men were appointed jointly in February, 1794; less than two months later Moorcroft resigned. His reasons have never been fully established, and were probably complex. He had no desire to leave his Oxford Street practice which was deservedly popular and financially rewarding. He can have had no illusions about cooperation with Coleman, whose personal arrogance and poor, virtually non-existent, knowledge of veterinary medicine did not augur well for the future teaching at the school. He must also have been aware of the desperate financial straits of the college, which by this time was all but bankrupt. He may even have had in mind a wish for freedom to pursue wider horizons, as he was eventually to do in his travels in remote regions of the Himalayas and Kashmir, where he ended his life in what is now Uzbekistan, killed by native tribesmen, in 1825. Moorcroft's biographer, Garry Alder, describes him as temperamentally restless and ill-fitted to work within the confines of an institution.

Whatever the immediate reasons may have been for the resignation, it left Edward Coleman (1765–1839) (Fig. 14) the only contender. He was later to protest that he felt himself 'ill-used' by Moorcroft's sudden withdrawal. He also admitted that he 'felt it

EDWARD COLEMAN, ESQ.ᴿ
Professor at the Veterinary College.

Fig. 14. Edward Coleman (1765–1839); engraving by Read after marble bust by Sievier (courtesy Wellcome Institute Library, London).

would be presumptuous, perhaps dishonourable for me so little versed in veterinary matters, to superintend the interests and growth of the infant school'[19]. In the event he appears to have had little difficulty in overcoming his qualms. He ran the school, to nobody's advantage save his own, and with a minimum of background knowledge, for nearly fifty years until his death in 1839. On his appointment, and when the school's ailing finances had been rescued by a parliamentary grant[20], one of his first acts was to shorten the students' course drastically, ostensibly to hasten the production of veterinary surgeons needed in the war with France which had begun in February, 1793. When the emergency was over, the duration of the course remained a few months instead of the three years originally planned by Sainbel and Granville Penn.

The inadequacy of a course of such shortness becomes all the more obvious when it is compared with teaching in other European schools at the time. The Swedish school at Skara offered a course lasting a minimum of two years; in Vienna the extensive course also lasted at least two years; in Copenhagen the students were expected to stay for three years. The France of Napoleon's expanding empire boasted a total of five veterinary schools: the original two at Lyons and at Alfort, plus the Turin school in Piedmont, one at Zutphen in the Netherlands, and lastly one at Aix-la-Chapelle, today Aachen in Germany, Charlemagne's old capital. For these five schools a decree of January, 1813, established a two-tier educational system, aimed at creating two professional levels. Pupils completing a three-year course in any of these schools qualified as *maréchaux-vétérinaires*; from among these were chosen the ablest students to study at Alfort, for a total of five years, to become *médecins-vétérinaires*. At this time Alfort had a teaching staff of seven in addition to the resident director; the other schools had to be content with four '*professeurs*' each[21].

The shortened course was not the only point on which Edward Coleman compromised the early hopes for the London school. He never changed his firm opinion that animal medicine was inferior to its human counterpart, and that in any case the horse was the only species of domestic animal worthy of any attention at all. Even with regard to the equine species, his few publications bear witness to the depth of his ignorance of the subject he taught for so long. Although resistance to Coleman among members of the veterinary profession grew steadily during his years at the head of the College, he was protected by its Governors. They included a number of medical men sympathetic to Coleman, notably his close friend Sir Astley Cooper (1768–1841), and they seem never to have questioned his high opinion of himself, or to have interfered with his running of the school[22].

Among the many people Coleman offended and alienated by his high-handed incompetence during a long career, William Youatt (1776–1847) stands out as the man who, in other and happier circumstances, might have successfully led the British veterinary profession in the first half of the nineteenth century. Born the son of an Exeter surgeon, he was educated for the nonconformist ministry. In 1810 he came to London; a year later he left the ministry and joined Delabere Blaine (1768–1848) in his veterinary practice in Oxford Street. Blaine encouraged Youatt to obtain proper qualifications at the London College. This was a disinterested suggestion, since Blaine was among the many whose relations with Coleman were cool. Perhaps for this reason the outcome was a similar lack of sympathy between Blaine's protégé and the professor: Youatt left the school without a diploma. Over the next three and a half decades, Youatt's reputation rose to heights never reached by Coleman's. His publications, highly regarded and home and abroad, included treatises on health and disease in sheep, cattle, pigs and dogs as well as horses. With William Percivall (1792–1854) he edited *The Veterinarian*, founded in 1828 and continuing as a most respected professional journal until it ceased publication in 1902.

Youatt's volume on the dog includes a chapter on rabies in which he showed his breadth of vision and his willingness to consider all species of animals in relation to disease. Like Samuel Bardsley, Youatt believed explicitly in the transmissibility of rabies, and had no time for the advocates of spontaneous occurrence. He wrote:

> What is the cause of this fatal disease, that has so long occupied our attention? It is the saliva of a rabid animal received into a wound, or on an abraded surface. In horses, cattle, sheep, swine, and the human being, it is caused by inoculation alone; but, according to some persons, it is produced spontaneously in other animals.[23]

'Some persons' may have been an oblique reference to Coleman, who had been broadcasting the view that rabies was generated spontaneously in ill-ventilated kennels. Youatt also compared the relative susceptibility of different animal species, the dog topping the list with three out of four dogs bitten becoming rabid, whereas only one in four humans would develop the disease. Cattle and sheep, in that order, were even less at risk. He admitted to scant knowledge of the nature of the 'rabid virus', but considered the seat of the disease similar to that obtaining in cases of 'variola and the vaccine disease, glanders and farcy'. In each case the respective tissues became the 'depôt of the poison'. In the case of rabies, that 'depôt' is in the salivary glands: 'in them it is formed, or to them it is determined, and from them, and them alone, it is communicated to other animals'[24].

It is in this context that we must see Youatt's struggle to come to terms with the growing number of dogs kept in towns and cities. He was a warm advocate of kindness to animals and deplored the many cruel practices of his time, publishing a volume on *The obligation and extent of humanity to brutes*. Its chapters, replete with biblical quotes and theological references and arguments, reflect the influence of his early training for the ministry. Perhaps as an extension of this humane attitude, in a rabies-ridden century, he took up Bardsley's call for a mandatory eight months' quarantine for imported dogs, and also demanded a reduction in the number of 'useless and dangerous dogs' kept in town and country. Like Balzac *père* in France earlier in the century, Youatt was acutely aware of the social consequences, in a time of frequent and severe outbreaks of rabies, of the presence, especially in cities, of large numbers of unwanted and uncontrolled dogs. And like the elder Balzac he suggested the imposition of a dog tax: 'A tax should be laid on every useless dog, and doubly or trebly heavier than on the sporting dog. No dog except the shepherd's should be exempt from this tax,...'[25]. Unfortunately there were powerful interests working against Youatt's sensible ideas. Pet dogs were becoming a cherished part of the Victorian way of life. With increasing numbers of dog owners, and a growing industry catering for the needs of their pets, social and commercial pressures impeded attempts to introduce control by legislation for many years after Youatt's death[26].

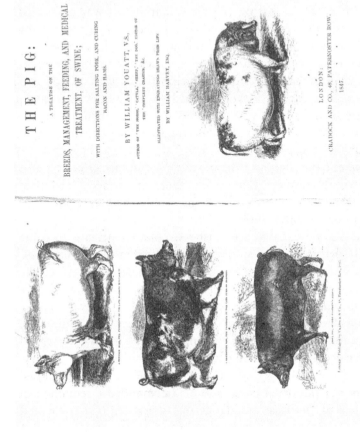

Fig. 15. Title page with main breeds of British swine in the nineteenth century, from William Youatt's *The Pig*, 1847 (courtesy Wellcome Institute Library, London).

Youatt was one British veterinarian who in the first half of the eighteenth century wrote treatises on a number of species of domestic animals (Fig. 15); but he never completed his formal education. His forthright honesty brought him into conflict, not only with Coleman, but also after Coleman's death with some fellow members of the profession who did not entirely share his idealistic views. His books on cattle, on sheep, and on swine were carefully compiled monographs on species ignored, in theory and in practice, at the London Veterinary College during Coleman's reign. In the preface and introduction to *Cattle*, Youatt made his feelings very clear. He referred to 'the absence of efficient instruction concerning the diseases of cattle in the principal veterinary school [William Dick had founded a rival school in Edinburgh in 1823], and the incomprehensible supineness of agricultural societies, and agriculturalists generally...', pointing out the importance of cattle, so often neglected in favour of the horse, for 'our agricultural prosperity and...the comforts and...very continuance of life'. Nevertheless his most original observations are to be found in his books on horses and on dogs. Writing on the horse he emphasised the danger, by then widely recognised, to which grooms and veterinary surgeons were exposed when attending glandered horses[27].

In spite of the differences between Youatt and factions with a more commercial outlook within the veterinary profession and agriculture in general, he was recognised in his time as representative of everything that was best in British veterinary science. In his volume on the dog he made a brief, prophetic observation, regretting a lost opportunity which came to fruition long after his death, in the work of Galtier and of Pasteur. He wrote:

> I very much regret that I never instituted a course of experiments on the production and treatment of rabies in [the rabbit]. It would have been attended with little expense or danger, and some important discoveries might have been made.

Thirty years later the rabbit, with its predominantly paralytic form of the disease, proved to be the ideal experimental animal for rabies research, and for the development of a vaccine.

In 1823 William Dick founded a Scottish counterpart to the

London veterinary school. He fiercely defended its independence until his death when, by his bequest, its ownership passed to the city of Edinburgh. Dick was a dedicated teacher who wrote a manual of veterinary science which Youatt called 'valuable', only to proceed to refer to the 'peculiar views' on rabies expressed by Dick. A Christian and a kindly man, Youatt gave Dick the benefit of doubt. The existence of a school north of the border did nothing to unify the profession, although a Royal Charter for the College of Veterinary Surgeons was, after many difficulties, obtained in 1844, only to arouse bitter renewed criticism from Dick[28]. The veterinary profession in Britain continued to be plagued by internal strife, whereas in contemporary France and Germany cooperation within the profession, and its easy acceptance by medical colleagues, had been building up to an amiable coexistence and a climate in which comparative medicine was flourishing. Youatt had been well aware of this, and his published works reflect awareness of the developments on the European continent.

When Coleman died in 1839 the school continued without interruption on its erratic way under the direction of the man who had for forty years been Coleman's uncomplaining lieutenant. William Sewell (1780–1853) had joined Coleman on receiving his diploma from the school in 1799; when he died more than fifty years later, he was principal of the school and president of the Royal College of Veterinary Surgeons. He was a conscientious, well-meaning Quaker who never showed any traces of the kind of constructive imagination that would have been needed to raise the college and the education it offered from its level of humdrum indifference in comparison with schools elsewhere. Fortunately, other forces were gradually coming into play. In addition to Sewell's position as *the* professor, the school now had an assistant professor in veterinary anatomy, and some chemistry and *materia medica* were being taught. Then, in 1842, James Beart Simonds (1810–1904) was appointed to a newly established chair of cattle pathology. Youatt and his co-editor of *The Veterinarian*, William Percivall, had long been campaigning for a broadening and lengthening of the students' course at the veterinary college; with Coleman's death the door was at last open for innovations. The all-important chair in cattle pathology was established with financial help from the Royal Agricultural Society, which had been founded the year before the death of Coleman[29].

When applauding the introduction of cattle and their diseases into the curriculum of London's veterinary college, it should be remembered that the new appointments only very belatedly brought the London college into line with the continental schools. The Paris school at Alfort had achieved comparable positions for its staff in 1782, well before the Revolution and nearly ten years before the London school came into existence, with its one professor and its heavy and long-lasting emphasis on the horse as the only species of interest. In 1828 France had even acquired a third veterinary school at Toulouse, to serve specifically the rich cattle lands of the south; it had long been demanded by societies concerned with agriculture and husbandry in south west France[30].

In Germany there was similar enthusiasm for veterinary subjects, as indeed there was in Italy; and in all of the major continental schools veterinary education had been more extensive, both in time and in number of subjects taught, throughout the period when the London school had languished under Coleman's misguidance. The superiority of the continental schools was reflected in the quality of the students they were able to attract, and in the resulting status of the profession. From the early years of the century, a growing number of medical students and graduates took advantage of the opportunity to acquire additional veterinary knowledge. It was a deliberate move to gain insight into a comparative approach to a wide range of new subjects, particularly in the fields of infectious and zoonotic diseases, which were rapidly become established as foci for new experimental research in the biological sciences.

It was a propitious time for James Beart Simonds to enter the academic field. A graduate of the London college, he had spent a dozen intervening years in practice where, first in his native Suffolk and subsequently at Twickenham west of London, he had had working experience of cattle diseases. His teaching introduced a new dimension to the curriculum of the school. His simultaneous appointment as consultant veterinary surgeon to the Royal Agricultural Society ensured that he in no way lost touch with practical work, and this is reflected in his publications. Whether he wrote on diseases of cattle or of sheep, or indeed on the teeth as a basis for determining the age of oxen, sheep and swine, his advice was always based on well observed fact. His evaluation of inoculation as a means of controlling bovine pleuropneumonia, carried out for the Royal Agricultural Society, rejected Dutch claims with detailed argument and careful experimental work[31].

During Simonds' time at the college, the veterinary profession had to face its sternest practical and political test. Since its ravages throughout Europe during the eighteenth century, rinderpest had assumed a relatively low profile, although reservoirs in parts of Russia continued to present a potential threat[32]. In 1859 Simonds could write that 'Rinderpest is a disease which specially belongs to the Steppes of Russia', and that '…in general terms, Rinderpest has not existed in Central and Western Europe for a period of forty-two years'. These were among his conclusions to a report written after his return from a journey undertaken to assess the danger of a general spread of the disease in Europe, and to the British Isles in particular, from outbreaks in eastern Europe and the Balkan States. Having ascertained that the disease was not present in areas from which cattle were imported into England, and that any progress would be hindered by 'hundreds of military cordons' before it could even reach the 'western side of the German states', he concluded that 'all alarm…may cease with reference to its introduction into the British Isles'[33]. He was soon to be proved wrong.

Varying opinions have been expressed concerning the relative merits of the views and attitudes of James Beart Simonds, and of John Gamgee (1831–94), when rinderpest finally reached London in June, 1865. The disease was introduced with cattle imported from the Baltic port of Revel, now Tallinn on the northern coast of Estonia (Fig. 16). Sir Frederick Smith, always given to strong opinions uncompromisingly expressed, in no uncertain terms blamed Simonds for his failure to anticipate the outbreak and to prevent the subsequent disastrous spread of the disease; for Gamgee he had only praise, and confidently believed that he alone was responsible for the eventual control of the outbreak. Pattison, on the other hand, concluded that 'Simonds deserves major credit for handling officialdom and eliminating cattle plague, and that John Gamgee played the minor role of publicist for control measures already advocated and fully understood by Simonds and others'. It is true that Simonds as early as 1848, when he wrote on sheep-pox and again in 1857 when he testified to a parliamentary select committee on rinderpest abroad, had shown good under-

ARRIVAL OF CATTLE AT THE RAILWAY TERMINUS, EUSTON-SQUARE.

Fig. 16. Circumstances favourable to the spread of contagion: cattle transport arriving at Euston Square railway terminus, 1849, from *Illustrated London News* (courtesy Wellcome Institute Library, London).

standing of the contagiousness of all these diseases, and of the value of quarantine as a means of control. But it is also true, as Sherwin Hall has pointed out, that Simonds failed to take the immediate action warranted by the situation when, as the government's chief veterinary inspector, he first came upon the initial cases in June, 1865, in Mrs Nicholl's dairy in Islington[34].

In 1865 John Gamgee was thirty-five years old and at the peak of his powers. In 1857 he had founded a veterinary school in Edinburgh, a gesture of defiance and challenge to William Dick at whose school he had taught for a year, during which time the two of them had found very little ground for agreement. Gamgee's venture proved to be a financial disaster, and by 1865 he was in the process of transferring his college to London, where it became the Royal Albert Veterinary College. It was to be as short-lived as his New Edinburgh school; and the arrival of rinderpest in Britain caught Gamgee, who had warned of precisely such a sequence of events, at a difficult time.

John Gamgee was a second generation veterinarian who had benefited from a cosmopolitan upbringing. He was one of three brothers who, in their various ways, left their mark on medicine and veterinary medicine in Britain in the nineteenth century. Their father was a veterinarian who lived in Florence when his children were born; like many a man of integrity of his time he had no illusions about Coleman[35]. His eldest son, also Joseph (1828–95), began studying his father's subject, but switched to medicine; John, on the other hand, having intended studying medicine, changed to veterinary medicine partly out of filial piety. The youngest brother, Arthur, was an early physiological chemist. John Gamgee was the only British veterinarian of his time with a sufficiently broad outlook to be a leading light in the formation of the first European veterinary congress held at Hamburg in June, 1863. The vast majority of participants was German, or German-speaking from Switzerland and Austria; there were a few Scandinavian representatives, but no French names can be found among the many contributors to the discussions. Only one Englishman kept Gamgee company: William Field, vice-president of the Royal Veterinary Society[36].

We have only Gamgee's own word for his central rôle in the organisation of the congress, and no way of knowing whether he exaggerated; but the depth of the discussions, especially with regard to rinderpest and its prevention in western Europe, is evidence that Gamgee's opinions might have indicated the road to preventing the cattle plague from reaching English ports in 1865, had the authorities been prepared to listen. They did not listen, and Simonds does seem to have hesitated too long over the initial outbreak. The result was disaster on an almost unprecedented scale. It further dented the already frail reputation of the veterinary profession in the minds of both the public and of the scientific community in Britain at a time when, in France and in Germany, veterinary science was going from strength to strength.

John Gamgee left the following year for America. The family biographer tells us that he went 'on the invitation of the United States' Government to undertake the study of Texas fever', but offers no clues to the initiative for this move. It was certainly a welcome opportunity, since his Albert College was about to succumb to final bankruptcy. He spent the next eight years in the United States working on reports for the Department of Agriculture, and for the Chicago Pork Packers' Association, on cattle diseases, on the effects of various methods of transport on the quality of beef cattle, and on the possibilities of improving refrigeration of meat. The latter offered scope for an early interest in engineering which Gamgee had forsaken to go into the veterinary profession; but it was hardly a fitting conclusion to a life which had seemed at one time to offer such promise for veterinary education and for the profession in general. Even in the European context Gamgee had at one time been in the forefront in a manner unheard of among British veterinarians. Brought up with a working knowledge of the major European languages he had been in his element in Hamburg in 1863.

There were a number of reasons for the failure of the authorities to arrest the spread of cattle plague in the 1865–6 outbreak. From the beginning successive Orders in Council were issued, none of them effective. Commercial interests were strong, and were strongly opposed to import controls and even more to slaughter on a massive scale which alone might have prevented the extent of the ensuing disaster. The vested interests of the country's cattle dealers gained the support of farmers, who had been promised no compensation for animals ordered to be slaughtered by inspectors

INSPECTION OF FOREIGN CATTLE AT THE METROPOLITAN CATTLE MARKET

Fig. 17. Too late: inspection of foreign cattle at the Metropolitan Cattle Market, 1865. From the *Illustrated London News* (courtesy Wellcome Institute Library, London).

sent in by local authorities (Fig. 17). The policies adopted were remarkably short-sighted and very inferior to those drawn up by Thomas Bates and the London Justices of the Peace 150 years earlier. The devastating losses resulting from the adoption of mercenary and foolhardy policies before 1865 were forcefully described by George Fleming five years later when he wrote:

> ...The losses from only two exotic bovine maladies ('contagious pleuro-pneumonia' and the so-called 'foot and mouth disease') have been estimated to amount, during the 30 years that have elapsed since our ports were thrown open to foreign cattle, to 5,549,780 head, roughly valued at £83,616,854. The late invasion of 'Cattle Plague', which was suppressed within two years of its introduction, has been calculated to have caused a money loss of from 5 to 8 millions of pounds...'[37].

The many unfortunate aspects of this outbreak, and the failure to control it, did nothing for public confidence, or for the morale of a veterinary profession still struggling to overcome the disadvantages of troubled beginnings and of continuing difficulties with its structure and education.

The difficulties remained throughout the 1860s and 1870s, as the veterinary schools in London and in Scotland attempted to come to terms with each other, and with the authority of the Royal College of Veterinary Surgeons. The diploma of the latter was finally approved as the sole entitlement to inclusion in the profession of veterinary surgeons by the Veterinary Surgeons Bill, which became law in August, 1881. Shortly afterwards the RCVS, although still denied a government grant, at last found more suitable headquarters in London[38]. The two events signalled an end to the overwhelming preoccupation with politics and educational status which had plagued and inhibited the profession for decades: crucial decades when their colleagues on the continent had joined medical scientists in the developments leading to the emergence of new disciplines in the second half of the nineteenth century.

It was not until the last decade of the century that the British veterinary profession began to approach the standards in comparative medicine which had long been established among their continental neighbours. Even resident veterinary surgeons at the Brown Institution (chapter 10) had remained relatively obscure and often unacknowledged when they played an active part in research conducted by their medical superiors at the institution. In 1880 a conference on animal vaccination held in London gave George Fleming (1831–1901), veterinary inspector to the War Office and author of a volume comparing rabies in animals and hydrophobia in man, an opportunity to air his views on comparative pathology in the *Lancet*[39]. Fleming made a plea for the introduction of the subject into the medical curriculum, and for chairs of comparative pathology to be established in all British medical schools, as they already were in continental schools. A knowledge of animal diseases, wrote Fleming, would be of far more importance to surgeons and physicians than the zoology and comparative anatomy they were taught.

The desirable combination of medical and veterinary knowledge advocated by Fleming as a basis for comparative pathology, and already practised in France, in Germany, and in Italy by men such as Chauveau and Hertwig and their pupils, was by 1880 about to acquire its first, but no less formidable, representative in Britain. John McFadyean (1853–1941), son of a Scottish farmer, graduated from veterinary school in Edinburgh in 1876. It was the year of Robert Koch's first paper on the anthrax bacillus. Its significance, widely reported, was not lost on young McFadyean, who had found the teaching of pathology in his veterinary course somewhat lacking. In 1877 he entered medical school in Edinburgh to obtain additional qualifications in science and medicine. Supporting himself by teaching in the veterinary school, he needed five years to complete the course. Only then did he consider himself ready to enter the field of comparative pathology and bacteriology as practised elsewhere at the end of the nineteenth century. He succeeded so well that he both made a personal impact and also lifted British veterinary medicine and research to a level comparable with its counterparts abroad[40].

Before the end of the nineteenth century, McFadyean was making important contributions to the subject for which he had so carefully prepared himself. In 1892 he had moved from Edinburgh to become Dean of the Royal Veterinary College in London, and to take up a new chair of pathology and bacteriology endowed by the Royal Agricultural Society. Two years later he became principal of the college. In that year of 1894, in his inaugural address at the beginning of the autumn term, he emphasised that the school was about to embark on a new era under his leadership. From now on, he told his audience, students would qualify for admission only by criteria comparable to those required for students of law or medicine. Their professional training would be achieved in a four-year course. Such an education would enhance their 'social and scientific' status; they would at last be able to claim their rightful place among the learned professions[41].

If this was not quite the end, it was at least a beginning of an end to the attitudes which had plagued the relationship between the medical and veterinary professions in Britain for a century. That had been in part a continuation of the contempt of past centuries for the cow leech and the farrier; it had not been improved during Coleman's time at the London college, his predominantly medical outlook supported by his Governors. In 1835 Youatt had expressed concern at the lack of understanding with which physicians and surgeons treated veterinary practitioners and their work; and he had asserted that the time could not be far distant when '...the veterinary surgeon, by his contribution to that neglected but all important subject, the knowledge of comparative pathology' would be better able to gain the respect of his medical colleagues[42]. Now, sixty years later, the rest of McFadyean's inaugural address was devoted to the subject which was to occupy him for the remainder of his long and active life: contagious diseases of domesticated animals, the means to control them, and their relation to the diseases of man.

Until he retired in 1927, McFadyean was committed to research and to the affairs of the college; but he also found time to edit, and to write extensively for, the *Journal of comparative pathology and therapeutics*, which he had founded in 1888. Its first issue had appeared almost fifty years after the publication of Rayer's *Archives de médecine comparée*; but where the latter had ceased publication with its first and only volume, McFadyean's journal became a lasting forum for discussion of comparative pathology in its widest sense. It remains an indispensable part of current literature, with only a minor concession to changing times and *mores*: 'therapeutics' disappeared from its title in 1965.

In his research McFadyean covered a wide range of diseases of farm animals, and wrote numerous papers, some of them definitive studies. He also campaigned ceaselessly for better control measures, more especially for such diseases as are transmissible to man: glanders, anthrax, and tuberculosis in cattle. His uncompromising stand with regard to the latter disease led to his celebrated clash with Robert Koch in 1901, at a time when Koch had convinced himself, and quite a few others, of his infallibility[43]. Koch's pure culture method had led to the identification of a number of agents of infectious diseases, the tubercle bacillus among them; and he had discovered tuberculin, an important diagnostic tool, although early hopes for its use in therapy proved unfounded.

At an international congress on tuberculosis in London in July, 1901, Koch advanced his latest theories and observations. In experiments with young cattle, he had found them not susceptible to tubercle bacilli of human origin. He concluded that human and bovine tuberculosis were two distinct diseases. Although the possibility of transmission to man of bovine tuberculosis could not be subjected to experimental investigation for obvious reasons, Koch maintained that observation indicated that transmission to man – and even more importantly child – through infected milk and butter '... scarcely exceeded hereditary transmission', and that consequently special control measures, let alone legislation, need not be considered[44]. When McFadyean came to speak two days later, he did not hesitate to disagree with the great man. His own observations had led to a very different conclusion. The *Lancet* commented: 'Much interest was taken in this paper, as it was an open secret that it would be a direct answer to the views of Professor Koch'.

Koch's main argument turned on the low incidence of primary intestinal tuberculosis, a condition which might have been expected to occur frequently if infected milk were the source. McFadyean's results of *post mortem* examination of children told a different story. Primary infection, apparently through the intestine, had been found in 28 per cent of cases. He cited the depressing statistics that almost one third of Britain's milking cows were suffering from tuberculosis in one form or another, and two per cent of all cows were excreting tubercle bacilli in their milk from tuberculous udders[45]. His message was clear: contrary to

Koch's opinion, it was of paramount importance to legislate against the threat of infection through contaminated milk.

The previous year McFadyean had published a paper on a very different disease of domestic animals. The disease was African horse sickness. Working in London with blood from infected horses in Africa, brought in sealed tubes by a returning English veterinarian, McFadyean showed the agent to be filterable. The concept of 'filterable viruses' was then less than 10 years old; neither Koch, in whose laboratories the agent of foot-and-mouth disease had been characterised as 'filterable' in 1898, nor McFadyean, then believed filterable viruses to be anything more or less than very small bacteria. By 1908, however, McFadyean had noticed something else about them: their inability to grow on artificial media. He described them as 'ultravisible...obligatory parasites' in a three-part editorial article[46].

With the work of McFadyean, and of his students and followers, British veterinary science at last came into its own, and was able to join on equal terms in the advance of comparative medicine in the twentieth century.

CHAPTER 7

From transmissibility of rabies and glanders to the bacteridium of anthrax 1800–70

Four years into the nineteenth century the ideas voiced by John Hunter in 1793 were subjected to experimental treatment, and the results described in a small book published in Jena, by Georg Gottfried Zinke[1]. A local physician of whose life little is known, Zinke's experimentation was simple and clear-cut. His slim volume describing the experiments and his conclusions stands as the first record of rational transmission experiments with a viral disease. The rest of the book is devoted to descriptions of a number of traditional materia medica used for centuries in vain attempts to stave off rabies. Zinke had read Hunter's 1793 paper on the disease, for he referred to it in footnotes in the book; but he made no attempt to acknowledge any conceptual influence on the formulation of his own experiments. In these he used a small brush to transfer infected saliva from dog to dog, to cats, rabbits, and even fowl, and he reported positive results of all cases of inoculation described.

The one experiment suggested in 1793 which Zinke did not perform, i.e. the inoculation of a dog with saliva from a human case, was carried out in Paris ten years later by Magendie and Breschet, who may not have known of Hunter's paper, and who only published their results, independently, later[2]. The interest in rabies of François Magendie, pioneer neurophysiologist, at this stage points to a growing realisation of the neurotropic character of the unknown agent of the disease. Hunter's paper had warned that the experiments suggested might be 'both difficult and dangerous'. Magendie made no attempt to play down the dramatic aspects of his experience in rabies research. With Gallic enthusiasm he described in detail his dealings with rabid mastiffs in an establishment for fighting dogs in Paris. He emphasised that in order to handle such specimens he found it necessary to be accompanied by '...students known...for their courage, sang-froid and dexterity'.

By the second decade of the nineteenth century, when Magendie was experimenting with rabies, questions began to be asked concerning another, possibly zoonotic, disease. When the French professor Sainbel at London's recently established veterinary college died of an acute infection in August, 1793, it was described as a 'putrid fever' with buboes on the face and body. He was buried hurriedly and the attending doctors issued instructions that no one was to approach the body owing to its highly dangerous state. Sainbel's friend Bracy Clark disregarded the instructions in order to produce a death-mask, since disappeared, and suffered no ill effects. The burial in a lead coffin with an outer wooden shell, however, owed nothing to the fear of infection, even of plague, as claimed by some modern historians; on the contrary, and possibly with its roots in fear of corpse-snatching by dissectors[3], it was a mark of respect according to the conventions of the day, bestowed on this Frenchman, with the 'generosity of the English', as if 'he had been a native of Britain', in spite of the troubled relations existing at that time of revolution in France[4].

In the late nineteenth century, a reinterpretation of the known facts in the light of later knowledge changed the tentative diagnosis of Sainbel's disease; a case was made for identification of the illness as glanders, transmitted from glandered horses in the stables of the college[5]. Sainbel himself did not believe glanders to be transmissible by inoculation in animals, and almost certainly did not suspect that it might be transmissible to man in any circumstances; nor are there any suggestions in the extant literature until the end of the eighteenth century that it might be so, in spite of the many outbreaks with heavy losses among army horses recorded during past wars[6]. One can speculate on the reasons for this, but supporting evidence is unlikely to be found. The possibility of a change in the organism seems remote; in view of the capricious and unpredictable nature of transmission of glanders to man – the danger of infection appears to be greater in the laboratory than in the field or in the stables – cases could have been overlooked or ascribed to other reasons. When standards of hygiene were low and diagnosis frequently inaccurate, possibilities of confusion with staphylococcal and streptococcal septicaemia, gangrene and erysipelas would have been legion.

The year 1800 may not have been a watershed in the history of infectious diseases, but it did mark the slow beginnings of a new era, a time of growing understanding of the nature and causes of infections. A number of factors contributed to a change of direction in pathology. There was the development of improved techniques and of the new methodology; there was an accelerating evolution of rational scientific inquiry; and increasingly there was a deliberate use of animal experiments designed to illuminate the problem. Initially they were simple transmission experiments with the two zoonoses rabies and glanders; gradually, through the century, animal models became the focus of experiments more deliberately designed to increase understanding of disease processes in man. The period yielded a number of fundamental discoveries in the field of public health and hygiene, which provided the basis for the emergence of the definitive disciplines of microbiology and above all medical bacteriology in the second half of the century. The results of some early studies suggested to their authors the existence of a *contagium vivum*, to which there had been pointers in the previous century. Before 1850 a number of results obtained in different areas of research gave added impetus to such ideas. Prominent among them were the works on yeast and fermentation carried out in Paris by Cagniard-Latour and in Berlin by Theodor Schwann; and also the findings of Agostino Bassi on the muscardine disease of silkworms, which provided a solid foundation for the concept of parasitic micro-organisms of vegetable origin as agents of transmissible disease. This work was all published before 1840, as was Ehrenberg's treatise on infusoria[7]. In 1840 followed a theoretical counterpart, Henle's essay on miasma and contagia[8]. During the same period, the considerable forces of Rudolf Virchow, Max Pettenkofer, and others, showed a certain amount of sympathy for the cause of anti-contagionism. Many of their arguments had a strong political flavour. Their rightful compassion for the poor of the industrialised nations of the nineteenth century, many of whom were badly nourished and whose living conditions were marked by often appallingly low standards of hygiene, led these otherwise astute observers to be reluctant to acknowledge the possible rôle played by living micro-organisms in the spread of diseases such as cholera and typhoid[9].

Before these debates reached their climax, reports of a number of distressing cases propelled glanders into a position of prominence in discussion of medical topics in the early years of the century. In 1820 Jean Hameau (1779–1851), country practitioner in his native Gironde, reported at the *Société de Médecine* in Bordeaux observations made in the course of caring for a local veterinarian, who died after three months of severe illness. The man had been sent to investigate sickness among horses in the common pastures around Audenge, soon diagnosed as glanders. Questioned by Hameau, he admitted to having plunged his fingers into the nostrils of glandered horses to remove pus, and to having frequently neglected subsequent hand washing because no water was available. Hameau was in no doubt concerning the conclusion to be drawn: his patient had died of glanders, which hence was transmissible to man. At the time scant notice was taken of this observation, and it remained buried in the archives of the society in Bordeaux. Years later it was incorporated in Hameau's likewise neglected *Etude sur les virus*, first submitted to the society's archives in 1836 and only published four years before his death. It was rediscovered and republished in 1895, the year of Pasteur's death, after the schools of Pasteur and of Koch had confirmed many of its observations and premature conclusions. Still accepting miasmatic causes for the intermittent fevers, Hameau was convinced that the diseases he classed as contagious could be caused only by living agents, 'multiplying at the expense of those they attack'[10]. Paradoxically, Hameau himself died of general sepsis after a minor operation.

At the time of Hameau's first observations of glanders, a contemporary textbook published in Vienna carried a warning that '...when opening carcasses of horses suffering from glanders or farcy, the utmost care must be taken in case of accidents not to introduce any pus into any wound, as this could lead to the most melancholy consequences and even death'. This warning was quoted in 1812 in a paper by a Dr Lorin, surgeon to a French

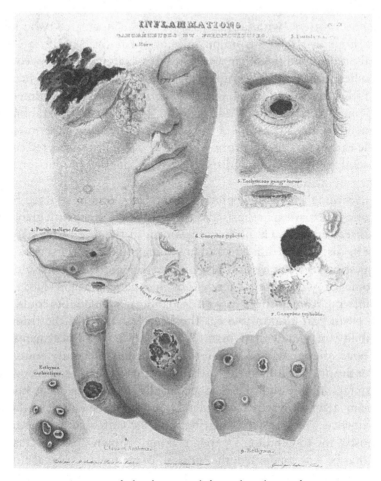

Fig. 18. Lesions of glanders (top left) and anthrax after Rayer, in his treatise on diseases of the skin, 1835 (courtesy Wellcome Institute Library, London).

regiment, who had observed the transmission of farcy to two members of the regiment, who had injured their hands while operating on horses with farcy. Lorin removed the small tumours (farcy buds) formed, and both patients recovered. A more detailed report of a fatal case in Berlin was described in 1821 by a Dr Schilling. His patient had been employed tending glandered horses; he became unwell but continued to work for six weeks until his illness became acute, when he died after a few days.

Schilling, bearing in mind the above textbook warning, carried out a post mortem 'with utmost care'. Then he inoculated two rabbits with material taken from pustules on the body. A week later one rabbit died and the other was killed. An examination showed both bodies to have pustules in various organs. There was no necrosis of the nose, and no further inoculations were performed. In Zinke's work on rabies, and Schilling's on glanders, the rabbit was introduced as a convenient subject for transmission experiments in the early years of a century which, toward its end, was to see that useful animal as an indispensable tool in the production of a vaccine against rabies.

Schilling's realistic depiction of the facial lesions suffered by his patient certainly emphasised the hazards of contact with glandered horses (Fig. 18). He himself remained, in the title of his paper as in its discussion, diffident and unwilling to take a firm stand on the question of transmission of glanders to man, wondering aloud if some of the horses attended by his patient might have been suffering from anthrax as well as glanders. Such doubts could have been in part responsible for the late realisation of the transmissibility of glanders to man. With other possibilities mentioned above, they remained well into the nineteenth century sources of confusion, including the possibility that disease following any contact injury might be, less specifically, what came to be known, from 1837 onwards, as pyaemia or septicaemia. Nevertheless, evidence was fast accumulating, and the editor added a warning postscript to Schilling's paper making his own position clear: as far as he was concerned, the 'poison' of glanders was transmissible to man[11].

In addition to such suggestive, but not conclusive, evidence from France and from Germany, more detailed observations were published in London in 1830. The author was John Elliotson (1791–1868), at the time a celebrated clinician at St Thomas's Hospital. A few years later he was to become uncritically enthusiastic about the practice of hypnotism and mesmerism, to such an extent that the issue clouded his mind and harmed his reputation. In 1830, in the wards of St Thomas's, two patients died within a couple of weeks in similar circumstances. In both cases there had been great prostration, abscesses on the extremities, pustules on the face, and apparent gangrene of the nose, with profuse discharge from the nostrils. Elliotson, a believer in contagion, searched in vain for a possible source of infection, until he read a report of the case of a corporal in the Dragoon Guards who had contracted fatal glanders the previous year from horses in his charge[12]. This case suggests that the possibility of transmission of glanders to man was in some quarters accepted by the public before it was recognised by the medical profession at large; indeed the concept of contagiousness was often more acceptable to the layman with common sense than to those with a large armour of rigid scientific principles to sustain them.

The report of the corporal's illness and death gave Elliotson the clue he needed. In a further search he found that as early as 1817 a student at London's Veterinary College had died in similar circumstances, after accidentally cutting himself while dissecting the head of a glandered horse[13]. Although an ass inoculated with material from an abscess on the patient's arm developed fatal glanders, the controversial Edward Coleman, head of the Veterinary College, and Benjamin Travers, F.R.S., surgeon at St Thomas's, agreed that the patient had died of 'constitutional irritation' rather than glanders, which they continued to insist could not be transmitted to man. Elliotson thought otherwise, and was supported by the *Lancet* which commented on the 'singular degree of blindness, or prejudice' of Travers in his dogged and outdated belief in Broussais's idea of causal 'irritation' as opposed to specific disease[14].

Elliotson's paper soon became an accepted point of reference for a number of related studies on the subject of glanders in man which appeared within the next decade. These studies formed part of a larger body of investigations of zoonotic disease, including rabies and anthrax, which laid the foundations for that confident comparative approach, which in its turn was to solve so many problems connected with infectious diseases in the latter half of the century. The very title of a paper published by K. H. Hertwig in 1834 illustrated the new attitudes. Hertwig's paper was called 'Transmission to man of animal infections', but the cases described were all of glanders or farcy in veterinary students, or in grooms employed in the stables of Berlin's veterinary school. Surprisingly he appears not to have made any attempt to inoculate animals with material from his patients; surprising because six years earlier he had attempted to transmit rabies from animal to animal by implantation of nervous tissue, although the results had been negative. Hertwig's (1798–1881) education had been deliberately tailored to the needs of comparative medicine. Having qualified in medicine at Breslau, he had travelled to major veterinary schools in search of additional knowledge[15].

Hertwig's career was symptomatic of changing attitudes on the European continent. Gone were the days when Bourgelat and his successors had been wary of 'debauched' medical students; nearly gone were the days when veterinary students saw medicine as a mere tool of social advancement. Animal experimentation and infectious disease research had come together, and veterinary science was helping to show the way to a new knowledge and better understanding of the nature, and of the pathways of transmission, of infections. Contributing factors were improvements and innovations in techniques, and from 1840 onwards in microscopes[16]. As noted elsewhere, attitudes changed far more rapidly on the European continent than in Britain for the better part of the nineteenth century.

In 1837 glanders was the subject of a monograph by P. F. O. Rayer (1793–1867), already known for his treatises on skin diseases and their classification. Rayer had a strong belief in comparative pathology as an essential tool for the study of all living beings. Without veterinary qualifications himself, he always made a point of consulting veterinarians when problems seemed to require it. In 1837, in the Charité Hospital in Paris, Rayer was unable to save a groom, who died after a short acute illness which appeared to have many features in common with glanders in the horse. Rayer had read Elliotson's paper; making enquiries he found that the groom had slept in the stable with a glandered mare before the onset of his illness. He introduced pustular matter from his patient into the nostrils of a sound horse, which duly developed typical glanders. Soon afterwards Rayer saw another case of human glanders in the same ward at the Charité; on the basis of these experiences he wrote the monograph on glanders and farcy in man. He found the disease a perfect object for his comparative interests; and three years later he compiled, with Gilbert Breschet (1783–1845), who had earlier worked with Magendie on rabies, a comparative study of glanders in man, equines, and other mammals. When this paper was presented at the *Académie des Sciences* in February, 1840, it led to a disagreement with Magendie who, his earlier rabies experiments notwithstanding, steadfastly refused to believe in the contagiousness of glanders in either horse or man. In the ensuing somewhat acrimonious discussion Breschet and Rayer were supported by the physicist Becquerel whose physician son, collaborating with a distinguished veterinarian, had made successful transmission experiments with the disease[17].

In the same year Rayer began publishing a journal which was a first of its kind, the *Archives de Médecine comparée*. Its appearance was premature, and it did not survive its first year. Much of the contents was contributed by Rayer himself and showed his wide range of interests, from recent outbreaks of foot-and-mouth disease to comparative studies of pulmonary tuberculosis in animals and man[18]. He distinguished clearly between nodules of tubercular and glanderous origin, which had been confused by veterinarians who regarded glanders as an equine form of tuberculosis. One of Rayer's pupils, J.-A. Villemin (1827–92) later in the century made definitive contributions to the study of tuberculosis[19]. The failure of the journal was a disappointment to Rayer, but his interest in the discipline remained undimmed. He was a founder member of the *Société de Biologie* in 1848 and its first president. From 1862 he was the first incumbent of the first chair of comparative medicine, established in Paris with the powerful support of Littré (1801–81). At the time Littré appealed successfully to the chauvinism and ambition of Louis-Napoleon to achieve this pioneering advance for France, before similar steps were taken elsewhere in Europe. A letter to the Emperor (Napoleon III since 1852), composed by Littré and signed by the Minister for Education, outlined with Gallic elegance and volubility the virtues of both research and of practical achievement in comparative medicine. It instanced the advantages of deliberate transmission of cowpox to man, and compared them to the sorry effects of accidental transmission of glanders, rabies, and anthrax. The letter also extolled the abilities of those already working in this field in France, and implored the Emperor to establish a chair in comparative medicine and to 'take an initiative which it is important not to leave to foreign schools'. In the decade leading up to the Franco-Prussian war, it must be a fair assumption that the 'foreign' schools uppermost in the minds of the letter-writers were German ones. At the end of the letter, Rayer was mentioned as the obvious choice for such a chair, since he had already proved his eminent worth with a number of comparative studies[20].

More than ten years before, Rayer had worked on another of the zoonoses mentioned in the letter to the Emperor. It was a disease which was rapidly becoming a main protagonist of the developments leading to the emergence of bacteriology as a scientific discipline during this second half of the century: anthrax. During the 1850s and 1860s, the loss of human life suffered in recent cholera outbreaks dominated the thoughts of those concerned with the developing public health movement. The Public Health Act of 1848 was followed by the formation in London of the Epidemiological Society in 1850. The name initially proposed for the society, the 'Asiatic Cholera Medical Society', leaves little room for doubt regarding its primary concerns[21]. On the other hand, for those attempting to use the study of animal disease to throw light on possible agents of infectious diseases of animals and man, anthrax presented a natural focus.

Like glanders, anthrax had frequently been confused with other complaints. Its transmissibility to man had been recognised in 1769 by Jean Fournier, a practising physician and medical officer in the Duchy of Burgundy, who wrote a number of texts on epidemic diseases of animals and man[22]. In his observations on anthrax, Fournier compared mortality in outbreaks of the disease in sheep with that in similar outbreaks of sheep pox, and wrongly identified several quite disparate lesions as different manifestations of '*charbon malin*'. But Fournier also had a social conscience, and he noted an increasing risk to man, associated with growing trade and developing manufacturing industries. He pointed out the hazards of exposure to artisans, to farmers, and to the poor, especially in the provinces of Languedoc and Provence. He referred angrily to 'infamously mercenary' shepherds and butchers, unscrupulously selling contaminated meat from animals dead of anthrax at reduced prices to the unsuspecting poor, who little realised that '...in this meat lay hidden the agents of their deaths'. Fournier went on to record, for the first time, cases of industrial anthrax, seen most frequently in the textile factories of Montpellier, which produced woollen blankets for a growing market at home and abroad. A century later, in the industrial north of England, this form of the disease was to acquire the sobriquet 'woolsorters' disease'[23].

If Fournier had had his problems distinguishing between lesions in anthrax and those due to other diseases, he handled concepts of contagion and of infectivity with confidence, in the language of his time. He referred to the agents of anthrax, and of other 'pestilential' diseases, as corpuscles, leaven, or molecules, of contagion. He vividly described the putative path of infection, when contaminated fleeces transmitted their 'atoms of anthrax ferment' to the unfortunate workers preparing, washing, and carding the wool, who were thus exposed to the immediate effect of 'these contagious corpuscles'. He also warned that such wool, however well washed, dried, and exposed to fresh air, could still retain for a very long time 'contagious molecules', able eventually to 'develop, multiply and reproduce with incredible speed'[24].

A decade later Philibert Chabert (1737–1814) wrote a treatise on anthrax in animals, which effectively removed a number of uncertainties. Chabert had risen above his origins as farrier and pupil of the elder Lafosse to become an instructor in Bourgelat's schools, and eventually to succeed Bourgelat as director of Alfort, in spite of his want of formal scientific training. In 1774 Chabert had been delegated to write, on behalf of the Alfort school, directions for the control of an epizootic outbreak of disease on the island of Saint-Domingue, better known to posterity as Haiti. The four-page memorandum adhered strictly to accepted practices for dealing with epidemics and epizootics. The remedies recommended, curative or prophylactic, relied greatly on vinegar and juniper berries, garlic and angelica root, camphor and honey. The importance of cleaning and fumigating stables, and of using clean water, was emphasised, since stagnant water was known to contain '…multitudes of animalcules and their eggs, which are easily passed on'. Chabert emphasised the difficulty of attempting a diagnosis of the disease from afar, but felt justified in assuming it to be a 'pestilential fever' complicated by large numbers of worms seen in dead animals at the scene[25]. In his conclusion, addressed to the local authorities, Chabert did not miss his opportunity to point out the advantages to an island 'so dependent for its riches on its domestic animals' of employing a qualified veterinarian, preferably 'a pupil trained at one of our schools'.

Writing on anthrax a few years later, Chabert differentiated between different manifestations of the disease in different species of domestic animals, distinguishing between equines, cattle, and sheep, and he emphasised the greater risk of exposure to animals compared with humans[26]. Few further facts were to be added to the literature until after the turbulent years of the Revolution and the Napoleonic Wars, when in any case the political emphasis on administrative aspects of medical practice, education, and public health predominated[27].

In 1823, eight years after the battle of Waterloo, Eloy Barthélemy, later the first veterinarian to act as president of the Academy of Sciences, demonstrated at Alfort the transmissibility of anthrax by feeding infected blood to horses[28]. At the same time, a spate of unfortunate accidents involving veterinarians indicated that man could be infected through accidental injury, when working with infected animals. The nature of the agent transmitted remained unknown and subject to speculation, as when Linnaeus described a disease with manifestations similar to those of anthrax as caused by the 'worm' *Furia infernalis*: a filiform, ciliate creature with 'reflected prickles' lurking on grasses in marshy areas or carried on the wind, to penetrate 'exposed parts of men and horses', with fatal results[29].

The question of priority for the first observation of anthrax bacilli has been the cause of some controversy between French and German writers over the years. The well documented facts of the observations following closely upon each other neatly serve to demonstrate the intense general interest, in mid-nineteenth century, in anthrax and in the potential of improved microscopes; an interest which enabled physicians and veterinarians in France and in Germany to observe certain small structures in the blood of animals dead of anthrax[30].

The observation most frequently cited in historical paragraphs of modern volumes of bacteriology and microbiology is that reported by Rayer in 1850. At the time, Rayer was working with his younger associate Casimir Davaine (1812–82), who in the 1860s went on to make many basic observations essential for the understanding of the aetiology of anthrax. In 1850 they were experimenting with anthrax blood, inoculating sheep in the laboratory and comparing the blood of animals dead from natural and from inoculated anthrax. They were primarily interested in the properties of the blood, distinguishing between the discrete corpuscles in healthy blood and those in anthrax blood, which were agglutinated in irregular formations. In his short report Rayer added a throw-away remark, which has become famous in the annals of bacteriology:

> The blood also contained small rod-shaped bodies [*petits corps filiformes*], approximately twice the length of a blood corpuscle. These small bodies showed no spontaneous movement.

Lagrange commented in 1938: 'Rayer, that encyclopaedist of comparative pathology saw and yet understood nothing. *Oculos habent et non videbunt*'[31].

The paper nevertheless remains the first published observation of anthrax rods. Five years later the German small-town physician Aloys Pollender (1800–79) published his own observations, which he reported were made in 1849 (hence the later arguments over priority). Unlike Rayer, Pollender was deliberately looking for disease agents, and he described in far greater detail 'rod-shaped bodies' (*stabförmiger Körperchen*), which he found in blood from five cows dead of anthrax. They were, wrote Pollender, 'extremely small, solid-looking, not entirely transparent, of even thickness throughout their length, not at all curved or constricted but perfectly straight and unbranched'. He also asked himself whether they might be lower animals or plants, and whether they might 'constitute the infectious matter itself, or are they simply vehicles for this matter, or have they nothing to do with it? These are questions I cannot answer.'[32]

Two years later the same rods were seen by the German veterinarian Friedrich Brauell (1807–82), who held a chair in the veterinary school at Dorpat (Tartu), in Estonia. Brauell was a typical exponent of nineteenth century interest in comparative approaches, and had studied both medicine and veterinary medicine at institutions in Germany and Denmark throughout his formative years. In 1857 a technical assistant in the school's anatomy department died of anthrax contracted during autopsy of animals dead of the disease. Brauell found non-motile rods in venous blood taken from the patient, and subsequently in blood from sheep inoculated with his blood. This was the first observation of anthrax bacilli in human blood; but when it came to interpretation and theoretical considerations Brauell confused rather than illuminated the issue, identifying the rods with 'vibrios' from putrefying material. He also noticed that embryos from animals dead of anthrax presented no lesions, and that their blood was normal. Unlike Davaine in the following decade, he drew no conclusions from this fact[33].

The final observation of the rods, before Davaine returned to make the subject his own from 1863 onwards, was made in Paris in 1860 by the veterinarian Henri Mamer Onésime Delafond (1805–61), at a time when anthrax was decimating the horses of the *Compagnie des Petites-Voitures*. In the early 1840s Delafond, then under the influence of Broussais's doctrines, had written on the epidemiology of anthrax, ascribing it to humoral causes. By 1860, the year before his premature death, Delafond had rid himself of the received unwisdom of Broussais (1772–1838), and described the rods found in the blood of affected animals. He speculated on whether this 'pathological product' were the 'cause or effect of anthrax', and even attempted, unsuccessfully, to induce multiplication of the rods, which he considered to be living 'organic matter belonging to the vegetable kingdom'[34].

Such were the early observations when, in 1863, Davaine came to reconsider the rods he and Rayer had treated so lightly in 1850. He later recalled that the observation had stayed at the back of his mind until, in February 1861, Pasteur published his paper on butyric acid fermentation. The agents necessary for fermentation Pasteur called vibrions, and Davaine was struck by an analogy with the rods he had seen in anthrax blood, and was spurred on to fresh inoculation experiments. Very soon he was able to claim the rods, which he now called *bactéries*, as the organisms causing anthrax. In 1863 he wrote:

> This agent is visible and tangible, it is a living organism which develops and multiplies in the manner of living beings. By its presence and rapid multiplication in the blood it produces in the constitution of this fluid, doubtless in the manner of ferments, modifications which promptly kill the infected animal.

It was a declaration of intent as much as of observation; and Davaine was to spend the rest of the decade in an attempt to obtain proof of this putative aetiology of anthrax. Later in the same year he gave a complete description of the bodies in anthrax blood and proposed to re-name them *bactéridies*, a term still used in France. He also demonstrated the inaccuracy of the old name of 'splenic fever' (*sang de rate*) for the disease, showing the rods to be present in the blood throughout the body and not just in the spleen[35]. The unusual persistence of infectivity in dried anthrax blood observed by Davaine at this time was to be explained by Koch in the following decade as being due to spore formation.

In order to prove conclusively that the *bactéridies* were the sole cause of the disease, it would be necessary to separate them from the blood. Filter techniques were not yet adequate for such separation; Davaine used the biological filter available in the guinea pig placenta, and inoculated a pregnant female with anthrax blood. When the female died two days later, her blood and the placenta were teeming with anthrax rods; the blood of the foetus was clean. Four more guinea pigs were inoculated with blood from the placenta, three with blood from the foetus. Only the former died of anthrax. This was the first recorded bacteriological filter experiment. More recent research has shown that the guinea pig placenta is not in fact an impenetrable barrier, but can very occasionally be transgressed by anthrax bacilli. Hence Davaine's elegant experiment has not entirely stood the test of time. It nevertheless remains an important first step in the development of techniques of bacteriological filtration and the methodology which within the next two decades essentially helped to establish the science of bacteriology, and shortly afterwards to define a new class of disease agents, the filterable viruses[36].

Davaine continued throughout the 1860s to work on aspects of anthrax infection. In 1864 he collaborated with Raimbert to consolidate and extend Brauell's earlier results comparing human and guinea pig anthrax. Raimbert, considered an authority on the epidemiology of anthrax, had formerly believed in the putrid nature of the disease, and been duly criticised by Davaine. Now Raimbert excised a malignant pustule from a patient and passed the material on to Davaine, who found in it the characteristic anthrax rods. Inoculated into a guinea pig the pustule matter caused death from typical anthrax, and 'considerable numbers' of anthrax bacilli were found in the blood[37].

Increasingly, Davaine became convinced that the bacteridium was the sole cause of the disease. Absolute proof was, of course, lacking, before the advent of pure cultures, and Davaine's work was frequently attacked by adherents of spontaneous generation and by those who did not believe in the specificity of disease agents. Undeterred he emphasised in 1868 that he considered his results sufficient reason to conclude that '...the development of the *bactéridies* is the cause of anthrax'. At the end of this productive decade, Davaine turned his attention to the bluebottle (*Musca vomitoria* L.) and its possible rôle in the spread of the disease. Shortly before, Raimbert had inoculated guinea pigs with probosces, legs and wings of bluebottles which had been in contact with anthrax blood. His guinea pigs contracted anthrax. Davaine was able to confirm these results, remaining convinced that the main source of infection in domestic animals was contaminated blood in the dust of stables, sheep-runs and pens, and invading the host organism either by contact with cutaneous lesions, or by inoculation via stinging insects[38].

After 1870 Davaine's research activities turned in other directions. When considering his contributions, it should be kept in mind that he belonged to an era, and a breed, which was nearing its close by 1870. Davaine was, and remained all his life, a Paris practitioner whose research was strictly a spare time activity, and except for his collaboration with Rayer, it was carried out in private laboratories, with no academic connections[39]. After 1870, in France and elsewhere, bacteriology and medical microbiology increasingly became the concern of academic institutions and laboratories.

CHAPTER 8

Putrid intoxication, animate contagion, and early epidemiology

In the early 1870s Casimir Davaine began a study of a condition which had caused confusion in his own work on anthrax, and in studies of glanders and other infections, in the first half of the nineteenth century. It was to continue to cause difficulties in the arguments developing over specificity of disease agents. For centuries associated with putrefaction and putrid substances, sometimes described by the blanket term of wound infections, an ever present hazard from the fields of combat to the surgical wards, such conditions became known by the more specific terms of pyaemia and septicaemia in the late 1830s[1].

Putrefaction and putridity and, by derivation, 'putrid intoxication': the stem of the word is Latin, the English equivalent of rot and rotten perhaps less elegant, but equally descriptive. In whatever form, the term must have been part of man's vocabulary since he first began to take notice of his organic surroundings and their inevitable eventual decay. Putrefaction was ever perceived as detrimental to health. The idea of a causal relationship of putrefaction to disease in man and in animals was formed at an early stage, although the nature of the process remained obscure. Some authors posited the existence of a particular putrefying agent capable of inducing an accelerating change in the manner of fermentation, a popular term associated with the action of leaven in bread, in organic matter and in living organisms.

The first rational experiments carried out in an attempt to elucidate the subject began in 1808, four years after the publication of Zinke's work on rabies. Transmission experiments with animals again formed the basis for developments, and there appear to have been similarities in the thinking behind the experimentation. The author was Bernard Gaspard (1788–1871), a young physician in a small town in the Saône-et-Loire, whose experiments were carried out intermittently between 1808 and 1821. The animals used were mostly dogs; but Gaspard was careful to widen his range by the inclusion of the occasional fox and suckling pig and, more importantly, a young lamb, concluding in the latter case that 'in this type of experiment herbivores and carnivores react in identical ways'[2]. The experiments involved inoculations, sometimes of the same dog on several successive occasions, with putrid material or with vaccine lymph, blood, bile, human urine, and also some of the simpler known products of putrefaction such as carbon dioxide, ammonia, or hydrogen sulphide. As Zinke had done in his rabies experiments, and Bourgelat had recommended in the case of glanders, Gaspard in some cases administered doses of various putative remedies in order to test their possible prophylactic value. He recorded the manifestations which developed in each case and concluded that carbon dioxide and hydrogen sulphide were harmless, whereas ammonia produced symptoms not unlike those caused by injection of putrid fluid. On the basis of his experiments he warned his readers against the consumption of 'black gamy meat, stinking game-birds, putrid ragouts, and infected cheeses' enjoyed with such 'disgustingly epicurean' abandon at the best tables[3].

The theme of Gaspard's study was taken up by a number of other authors, among them Magendie. Magendie enthusiastically endorsed and elaborated Gaspard's results, attempting to relate them more closely to human medicine. He claimed to have found that the most deleterious effects were obtained by intravenous injection of suspensions of putrid fish, which caused symptoms 'analogous to typhus and yellow fever'[4]. A noteworthy aspect of Magendie's short paper was a simple experiment with far-reaching sequels in the future: it showed the putrid fish fluid to be less harmful after filtration through the primitive paper filters then available[5].

In the following decade Theodor Schwann (1810–82) was completing his years as a student of, and subsequently assistant to, Johannes Müller (1801–58) in Berlin. He was one of the school of young physicians who, inspired by Müller, were to shape the study of histology and physiology in nineteenth century Germany. Having discovered pepsin in 1836, Schwann turned to the problems of alcoholic fermentation, and of putrefaction, in a series of classic experiments which led him to propose a microbial origin for both processes[6]. Similar views were expressed simultaneously by Cagniard-Latour in Paris in his work on fermentation[7].

Also in France at this time Pierre Adolphe Piorry (1794–1879), whose special interests included the creation of large numbers of fanciful medical terms few of which have had any lasting impact, managed to add to the vocabulary two words which were to outlast all his other innovations. After 1837 'pyaemia', and later 'septicaemia', became standard terms for what had hitherto been described as 'putrid intoxication' of the blood in inoculated animals, and in accidental cases in man, where the condition was sometimes associated with suppurative lesions (*pyoémie*)[8].

In the following decade Rudolf Virchow (1821–1902) attempted to clarify the subject by distinguishing clinically between pyaemia and septicaemia. When in 1856 he published his classic study of thrombosis and embolism, he included experiments with dogs injected with putrid material, confirming the results of Gaspard and of Magendie. He now referred to Gaspard's 'famous experiments with putrid injection'[9]. His discussion of the work of other contemporary authors serves to illustrate the confusions surrounding the concepts of diathesis and dyscrasia, especially in the way they were perceived against a background of former humoral theories. Virchow also referred at some length to Sédillot's study of pyaemia, which also paid particular attention to this problem[10]. In the same year of 1856 P. L. Panum (1820–85) carried out a series of experiments remarkable for the accuracy of their measurements, and designed to demonstrate the relative effects of varying doses of putrid blood, flesh, brain substance and human faeces when injected into dogs[11]. With the exception of Gaspard's few experiments with fox, pig and sheep, all of these 'putrid intoxication' studies were carried out on dogs, deliberately chosen as models for the effects of putrid substances in man. Among the several authors, Gaspard was the only one who registered compassion for his 'unfortunate victims of experimentation'[12].

In spite of many determined efforts, little real progress ensued until the 1870s. In the meantime, the terms septicaemia and pyaemia continued to represent disease entities of doubtful identity and unknown aetiology, confusing to practician and theoretician alike. Davaine's work with animal models in 1872 made some contribution towards reality and quantification in approach. He studied the relative effects of putrid material by passage in rabbits, rats, and fowl. Rabbits were found to be the most sensitive; age also played a part, young individuals being less resistant than older ones. Putrid blood was shown to increase in virulence by passage in rabbits, although the virulence gradually disappeared from unused samples. Such results formed an important and necessary foundation for further research. The active element Davaine described as a 'ferment of putrefaction'; he believed that it might be only one among several in putrid blood, whose existence could be unequivocally proved only when methods became available for isolation of such 'ferments'. 'Perhaps it will become possible to find such methods' wrote Davaine in 1872[13]. Within the next ten years methods for isolation of individual disease agents by staining and pure culture were developed, chiefly through the work of Robert Koch (1843–1910).

Also in the early 1870s, important work on sepsis was done by Edwin Klebs, not with animal experimentation but as a direct result of observations made in more melancholy circumstances: a study of the pathology of gunshot wounds suffered by troops during the Franco-Prussian War of 1870–1. Born and raised in Königsberg (at present Kaliningrad) in East Prussia, a city unhappily caught up in the protracted wrangles between Prussia, Poland, and Russia, Klebs was always a difficult and contentious presence, albeit a stimulating one, on the scene of early bacteriology[14]. Paradoxically, as a Prussian, his inexhaustible compassion and care for French troops in an improvised field hospital outside Berne at the end of the war in 1871 was little short of heroic. Its scientific by-product, Klebs's study of the pathogenic micro-organisms found in material from his patients in the field hospital, was a pointer to the possibilities of identification of disease agents which were to be explored by other means in the following decades. During this period of development Klebs initiated more studies than he followed through; even his description of *Corynebacterium diphtheriae*, or the Klebs–Loeffler bacillus, in 1883, remained to be confirmed, and the organism positively identified, by Loeffler in pure culture later in the same year[15].

In the course of his study of gunshot wounds Klebs suggested the preparation of porous filters of white unbaked clay. He hoped such filters would be superior to existing ones for the separation of pathogenic bacteria from the fluids in which they were found. The resulting method developed by his assistants Tiegel and Zahn[16], connecting the unbaked clay cells to a Bunsen air pump, was used in studies of anthrax and gunshot wounds. In the latter Klebs found and described a number of different micro-organisms; but his studies were short on the patience and accuracy of later work by Pasteur and by Koch. Consequently Klebs felt justified in interpreting his observations as showing that all the different forms seen by him were no more than varieties of one pleomorphic organism, which he named *Microsporon septicum*[17]. With Tiegel, he also used the newly developed filters to study the infectivity of filtrates and residues from anthrax material. Here may be found the first suggestions that filtrates could be toxic although not capable of transmitting anthrax experimentally in series[18]. To such modest and possibly unwitting beginnings may be traced much later work involving filtration experiments.

Klebs's filtration experiments were matched by a number of suspension and sedimentation studies, designed to determine the nature of infectious agents. In 1872 Carl Joseph Eberth (1835–1926) mixed anthrax blood with water, and found that the active principle settled out of the suspension leaving the supernatant inactive[19]. A similar method had been used by Chauveau in 1868 in experiments on vaccinia and variola, and by Burdon Sanderson with rinderpest at the same time; their work will be discussed in the following chapter. Davaine at the end of his anthrax studies was convinced that the presence of the rod-shaped bacterium was the necessary and sufficient cause of the disease. Certainly his own work had come very close to proof of a bacterial aetiology, but there remained weak points in the evidence. The anthrax problem still was only one among many complicating the wider issue of the understanding of the aetiology of infectious diseases in general. The work of Pasteur, ultimate exponent of a comparative philosophy, on silk worm disease, and his definitive demolition of the belief in spontaneous generation in the 1860s[20], provided supporting evidence. So did the studies on the classification of bacteria by the botanist Ferdinand Cohn (1828–98)[21]. The careers of the contributors in this field in the second half of the nineteenth century illustrate the wealth of interests brought together by the new discoveries. Eberth was a pathologist from Virchow's school at Würzburg, whose long working life coincided with the emergence of bacteriology as a scientific discipline. He was to make many contributions to its comparative aspects, studying typhoid fever as well as septicaemia in rabbits, and pseudo-tuberculosis in guinea pigs[22]. Ferdinand Cohn, at the other extreme of the range of professions involved, was a botanist whose early interest in algae and fungi gravitated towards bacteria, which became his main interest. He had offered strong support to Pasteur in the battles of wills and words to disprove the theory of spontaneous generation. In 1876 he became the champion of the young Robert Koch and his work on anthrax which was to prove essential for further development[23].

In 1876 Koch was an unknown country practitioner in a small town in an oft-disputed border area which was then Posen in Prussia (now Poznan in Poland), where he had settled after returning from medical duties in the Franco-Prussian War. At home, under primitive conditions, Koch spent his spare time in experimentation, drawing on solid foundations laid during his student years at Göttingen and in Berlin. At Göttingen one of his teachers had been Jacob Henle (1809–85), whose theoretical work on contagion may have kindled Koch's early interests and led him in the direction of medical microbiology. Henle's essay on miasma and contagia was written in 1840, at the end of two years spent in Berlin working with Johannes Müller and Theodor Schwann. They were years when work on yeast and fermentation, and on putrefaction, suggested the existence of parasitic micro-organisms as agents of infectious disease as a logical extension. Reviewing the work of others, including his friend Schwann's on fermentation, and Agostino Bassi's on silkworm disease, Henle concluded in 1840: 'The material of contagions is not only an organic, but a *living* one, and is indeed endowed with a life of its own which is, in relation to the diseased body, a *parasitic organism*'[24]. His theories created no immediate excitement, but provoked the following comment in the *British and Foreign Medical Review*:

> We are rarely sanguine of finding much that is profitable in German
> works on pathology; but if there be a writer in that country from
> whom more than from any other we should anticipate solid
> excellence in whatever subject he undertakes, it is Dr. Henle, the
> laborious and clever assistant of Professor Müller...[25].

In line with his prejudice against German pathology the reviewer
expressed disappointment in the work as a whole, but singled out
for praise the essay on contagion. It was indeed a first attempt to
integrate work on the aetiologies of a number of infectious diseases
of animals and man, and to show that results obtained by various
authors could be explained by the putative existence of specific,
living microscopic agents, acting as parasitic invaders of the host
organism. Henle quoted work on the 'contagious exanthemata',
typhoid, influenza, dysentery, cholera, plague, and puerperal fever.
A contagious origin of the latter condition was already being
discussed in a number of places at this time, in the decades before
Semmelweis first introduced his version of antisepsis in the
obstetrical wards of Vienna (see also the work of Adam Neale
below).

Explaining to his readers the 'reasons which prove the individual
life of the contagions', Henle wrote:

> The ability to multiply by assimilating foreign materials is known
> to us only in living organic beings. No dead chemical substance, not
> even an organic one, multiplies at the expense of any other; when
> brought together, they always enter into combinations from which
> the original quantities of the materials acting upon each other may
> again be separated[26].

Henle's style was succinct except for the fact, pointed out by
George Rosen, that exact terms of reference were lacking for
organising the framework he was attempting to construct. It was
his arguments and their arrangement into a trenchant theory
which were very much Henle's own; the data he borrowed from
a number of other observers. He also considered what he termed
the 'better known epizootic diseases' such as rinderpest, sheep-
pox and anthrax, in a group together with the exanthemata,
influenza, etc., which he noted were 'said to arise miasmatically
and to become contagious in their further course'. This was a
common creed of the period; according to Henle those in another
category, which included syphilis, scabies and glanders, were

among diseases of man and animals occurring 'only through contagion and which, nowadays at least, are not seen to develop miasmatically'. On hydrophobia and its perennially puzzling origins his observations were based largely on Hertwig's results, and reflected the uncertainties of his time. He concluded that even when appearing 'spontaneously' in dogs, rabies was unlikely to have been produced by miasma, since it occurred only sporadically and in any season, and apparently became 'epidemic' only through contagion[27]. The belief in spontaneous occurrence of rabies in dogs was to persist in some quarters even into the 1930s, on the eve of a real understanding of the nature of the filterable viruses[28].

Henle's treatise can be seen as a distillation and elaboration of the thoughts of those who did not agree with the views of the anticontagionists, rather than a new departure. It is not difficult to see his views as forming an inspiring basis for Koch's own ideas and further enquiries when he became his pupil twenty years later. From the *contagium vivum* to the outlines, however vague, of proving its causal relationship to infectious disease, the ideas which were to provide the springboard for Koch's definitive work on pathogenic micro-organisms are all there. From its eighteenth century beginnings, a general acceptance of a belief in living agents of disease had been slow to establish itself, although there had been straws in the wind. Garrison's claim that Henle's essay on miasma and contagia is 'the first clear statement of the idea of a *contagium vivum* is certainly an exaggeration[29]. The eighteenth century writings of Cogrossi and of Plenciz made their point equally clearly, even if in the terminology of their times they referred to the agents as minute 'insects'[30].

The year before the appearance of Henle's essay Henry Holland in London, otherwise more noted for Royal service than for depth of thinking, had published his jottings on medical subjects. Still referring to living agents as 'insects', Holland included a chapter entitled 'The hypothesis of insect life as a cause of disease?', hedging his bets with a concluding question mark[31]. Even earlier, when Henle was still a student at Heidelberg, Adam Neale had published a volume attempting to explore the 'Linnean doctrine of animate contagions'. Neale was a colourful character, an Edinburgh graduate who had been physician to the forces in the

Peninsular War of 1808–14. He had recorded his observations during this campaign, and also later experiences at the British Embassy at Constantinople[32]. In 1820, he settled briefly in Cheltenham; within a few months he was forced to leave after producing a pamphlet casting doubt on the authenticity of the water served to visitors at the fashionable spa. Not until 1831, the year before his death, did he set out to evaluate the 'Linnean doctrine', and to draw his own conclusions.

Neale took as his point of reference the views of Linnaeus's associate Nyander, who in a thesis supervised by Linnaeus claimed that all contagious diseases were associated with eruptions, externally or internally. Quoting Kircher, Leeuwenhoek, and Rivinus, Nyander concluded that such diseases could be caused by 'very minute insects...such for instance as *Acari* of various species'[33]. In the case of the annual outbreaks of dysentery in Sweden, the disease 'most probably derives its origin from the self same *acari* lurking in acidulous beverages and which, by means of the necessaries, are propagated and thus give rise to what is called contagion'[34]. In a short chapter on puerperal fever, Neale quoted a report in the *Manchester Advertiser* of April, 1831, which makes it clear that the transmission of 'animate' contagion from the dissecting room to the delivery room was a generally accepted fact during a local outbreak, fifteen years before Semmelweis drew his even then disregarded conclusions in Vienna[35]. Neale came down wholeheartedly in favour of the 'intro-animate pathology' of Nyander, supporting his opinions in a number of chapters on individual diseases.

Neale established to his own satisfaction that most infectious disease agents were 'very minute insects', in some cases acari (generally favoured by Nyander), in others (e.g. dysentery) 'very minute species of mite', or water-mites. He freely drew comparisons with diseases of animals and plants, included whole chapters on cattle disease and vegetable galls, and even appended an illustration juxtaposing galls on a maple leaf with 'animal hydatids' from a treatise on tubercular diseases. His final conclusion stated categorically that in most cases death of humans, animals and plants was caused by 'parasitical insects and animalcules, which entering, nidifying in, and preying upon, the bodies, produce this effect – death'[36]. Neale also introduced the notion of larger insects effecting mechanical transmission of disease of both animals and man, in particular the 'murrain, or plague' of cattle, and the 'pestilence, or plague in the human race'; he believed that 'plague is disseminated and inoculated by the bites and stings of noxious insects'. His theory of the development of an epidemic was complex and had no regard for specificity. He wrote of epidemic disease beginning as:

> merely a malignant fever, arising from the inhaling of putrid miasmata, either from the malaria of marshes, or of dead animals, soon after which the noxious flies or insects being called into life, by the heat of the climate, commence biting or stinging the bodies of the sick, and thereafter attacking those of the sound, the disease becomes extended and communicated in a manner so mysterious, so frightful, and so rapid, that it acquires the name of the plague or pestilence[37].

It is perhaps ironic that Neale's old comrade-in-arms and superior in the military hospitals in Portugal during the Peninsular War, William Fergusson (1773–1846), should have tended to an entirely different view of contagion: a view linking him to the anti-contagionists at home and abroad, and the ideas expressed forcefully by Nicolas Chervin in France at the time[38]. Retired from the services in 1817 and settled into a comfortable and lucrative practice at Windsor, Fergusson set down his views in a number of papers on individual diseases. In all of them he objected to what he considered the prevailing views of contagion, and the consequent quarantine laws. His arguments were in part sociological, since he believed as did many physicians at the time that fear, poverty, and general misery were the most important factors for the generation and spread of such diseases as dysentery and typhus. Compassion also led him to deny the usefulness of quarantine and isolation even in the more obvious cases of plague and smallpox. He argued that in epidemic situations these diseases were known to have penetrated the 'guarded gates of the Royal Palace', and had been able to surmount ramparts and make their way over and through stone walls: hence they must 'be considered as much atmospheric as contagious'[39]. Well in advance of Virchow's social document on the typhus decimating a community of poor and disadvantaged weavers in Upper Silesia in 1848, Fergusson wrote an essay on the disease designating it an 'atmospherical contagion of locality'. He noted its tendency to prevail in the 'wet and cold weather' of the winter months, and continued: 'Thus, the humid, ill-ventilated, and imperfectly heated dwellings of the poor are in such seasons its constant abode; and if to these we add the adjuncts of cold and fatigue, and sorrow and hunger, the sad concomitants of poverty, we need not wonder when we see it devastate the hovel and the cottage. Even so trifling a cause as the continuance of wet feet,...'[40]. The behaviour of cholera he also found puzzling. He wrote of 'the inexplicable phenomena of this strange pestilence'. Given the various apparently contradictory facts of its spread, he considered current quarantine laws '...in many instances...unnecessary cruel and mischievous'; those designed to keep out yellow fever he regarded frankly as 'eminent for absurdity'[41]. These were arguments, and attitudes to quarantine laws, widely accepted in the nineteenth century. The puzzling aspects of yellow fever epidemiology in particular could be, and were, used to advantage by factions opposed to quarantine legislation, many of whom had vested interests in overseas trade and the necessity of keeping open ports and trade routes[42].

The theme of social deprivation as a factor in the development of epidemics was taken up by William Farr (1807–83) in his annual reports when he became the first medical statistician, or 'compiler of abstracts', in the office of Britain's first Registrar-General of Births, Deaths, and Marriages. Certainly Farr's work helped to counter the criticisms of some of his contemporaries that medical statistics was a 'cold-hearted' study, treating patients as 'mere abstractions, or even as "ciphers", and not human beings...'[43]. His evaluation of the mortality in different London districts in the influenza epidemic of 1847 gave evidence of his concern for the effects of poor living conditions in the London slums. He found the mortality in that epidemic to be almost twice as high in the 'unhealthiest district of London' (St George in the East), as in the 'least unhealthy districts of London' (Lewisham)[44]. In his examination of annual epidemics of influenza and of smallpox, Farr was actively searching for mathematical expressions of the laws of epidemics; although he was familiar with the theories put forward by Holland and by Henle, such as they were, he found them interesting, but not especially helpful for his purposes.

Consideration of theories of 'epidemic infusoria' and specific animal contagion was all very well; unless and until the 'infusoria' had actually been observed under the microscope, their existence must remain a matter for speculation[45].

By the time he wrote his 30th annual report from the office of the Registrar-General, Farr could clarify his attitude to measures of quarantine and medical policing. By this time he had come to accept the importance of controlling 'germs' of 'zymotic diseases', from syphilis to smallpox, measles, and scarlet fever. His proposed control measures included enforcement of regulations governing assemblages of 'large masses of men in pilgrimages', and 'strict sanitary regulation' of vessels plying trade with distant shores. On the other hand, his scepticism regarding quarantine measures remained unabated. He wrote: 'The interception of the intercourse and commerce of nations by quarantine is injurious to their vital interests. It should be kept within the narrowest limits; and England should carefully abstain from treading in the steps of the fanatical populations of the Mediterranean'. Having earlier quoted Charles Darwin on the diseases of plants and animals under domestication, Farr concluded: 'As zymotic diseases of domestic animals are governed by the same general laws as the corresponding diseases of men, similar methods should be pursued in dealing with livestock'[46]. Applying a mathematical law tentatively based on the course of an epidemic of smallpox recorded in 1840, Farr was able, during the rinderpest outbreak of 1865–6, to predict its further course and eventual decline with tolerable accuracy in the circumstances. Farr began his attempt at prediction approximately half-way through the outbreak; the epidemic reached its peak a fortnight later than he had predicted. The table below shows the discrepancies between the values calculated by Farr and the actual figures for new cases registered, at four-weekly intervals. Explaining the 'obvious' nature of a curve based on these values, its initial rapid ascent slowing before reaching a maximum from which it would descend more rapidly than it had mounted, Farr attempted to allay public fears concerning the uncontrollable nature of the epizootic in a letter to the *Daily News* in February, 1866[47].

In his recommendations for measures to limit the extent and effects of epidemics, Farr had quoted a number of precautions to

Period ending	Calculated values	Actual figures
4 November 1865	9597	9597
2 December	18817	18817
30 December	33835	33835
27 January	47191	47287
24 February	43182	57004
24 March	21927	27958
21 April	5226	15856
19 May	494	14734
16 June	16	5000 (approx.)

be taken by those caring for patients with transmissible diseases, laid down by William Budd, his contemporary and fellow epidemiologist. Budd (1811–80), however, approached epidemiology from an angle essentially different from that of Farr. He wrote:

> Statistics afford much important information on epidemics; but it is all of a general kind. The really vital questions they leave almost untouched. Neither on the mode of being of the morbific agent without the body, nor on its mode of action within, do they throw any but a dim and distant light. Happily, there is another principle at our service which is of far greater promise[48].

To Budd, the key word and concept was 'contagion': 'In the act of contagion, we are brought into direct relation with the epidemic poison at one very important phase of its existence'.

In the same year of 1863 B. W. Richardson, addressing the Epidemiological Society, warned that: '...statistics...have been carried in respect to epidemics to the extremity of their present application'. Instead, in view of the well recognised specificity of 'epidemic disorders...in particular classes of animals', important lessons could be learned from studying such disorders in domestic animals[49]. This was a theory which Budd was already putting into practice. Since August, 1862, he had been a member of a committee appointed by the British Medical Association to investigate epidemic and epizootic diseases[50]. His writings on anthrax, on sheep pox, and on the committee's work on epidemic and epizootic diseases all reflect his belief that advance towards successful control of epidemic diseases in man could be achieved only through study of epizootic diseases of domestic animals. Writing on sheep pox, he emphasised that only in domestic animals could contagious diseases be studied under properly controlled conditions: all affected animals could be kept under observation simultaneously; their movements and mutual intercourse could be controlled; both natural and inoculated disease could be studied experimentally[51]. In this study of an animal disease, Budd acknowledged his not inconsiderable debt to Beart Simonds, professor of cattle diseases at the Royal Veterinary School since 1842. Simonds had, 'with the liberality of a true man of science', offered him unstinted help in theory and in practice[52]. Something of a loner within his profession, Budd in this paper acknowledged his dependence on veterinary advice rather more freely than was customary among his medical colleagues.

When in the early 1860s Budd drew attention to anthrax in the pages of the *Lancet* and of the *British Medical Journal*, it was regarded in Britain as a disease of sheep and cattle; transmission to man was rarely if ever described in the literature, although frequently discussed on the European continent[53]. Quite apart from the experimental work pursued at this time in France and in Germany, interest in the disease was also rekindled by accidental transmission of malignant pustules to man from horsehair, newly imported from Russia and Brazil, and used especially for upholstery in the manufacture of furniture. It was a source of anthrax to rival in importance the wool indicated by Fournier in the French blanket factories in the later eighteenth century; within a short time, it was also revealed as the source of 'woolsorters' disease', again in imported wool and hair from Russia, in the North of England in the 1870s and 1880s[54]. By then, the aetiology of anthrax was no longer a subject for speculation. The identity of its agent, its behaviour in pure culture, and its different manifestations in man and in animals, had become part of the known facts of a well-established germ theory, which was developing into the new bacteriology.

The gap between Farr's statistician's approach and Budd's contagionist's views never prevented their mutual respect. In the course of his work on successive cholera epidemics, Farr also gradually changed his attitude to fall into step with contemporary developments in epidemiology based increasingly on advances in microscopy and pathology. In his report on the cholera epidemic of 1866, he was ready to acknowledge the importance of identification of 'elementary disease particles' seen under the microscope[55]. By then, the contagionists' view of epidemiology had received a boost from study of a totally unrelated disease: the outbreak of rinderpest which culminated in early 1866, and the decline of which Farr himself had so successfully predicted.

At the end of that outbreak, in July, 1866, Burdon Sanderson, as member of the Royal Commission concerned with the epizootic, summed up the lessons in epidemiology learned during the fight to control the disease[56]. Like Budd earlier in the decade, he emphasised the advantages of studying transmissible disease in animals; and as Layard had done in 1757 (chapter 4, note 20), he drew attention to the psychological advantages of observing and working with animals: 'In man, pain, weariness, irritation of temper, or, as it is called, worry, and other purely psychical conditions, interfere both with the pulse and respiration. Bovine animals are never nervous or hysterical'[57]. He also noted that such studies were pursued with far more enthusiasm on the European continent than in his own country.

It is in fact clear from Burdon Sanderson's short paper that, apart from a few general introductory remarks aimed at placing epidemiology in context in its relation to the practice of medicine, his message is that of the comparative pathologist he was to show himself within the next few years. He was firm in his belief that further progress in epidemiology could now be made only when the nature of infectious agents had been explored and determined. Diffusion experiments which had been carried out by Chauveau in France, and by Burdon Sanderson himself with 'the virus of cattle-plague', suggested that infectious agents were associated with 'colloid' rather than 'crystallisable' constituents of infected blood. In 1866 Burdon Sanderson modestly called this an 'inconsiderable step towards separation of the virus'. Within a few years he would be expanding his research into this and other subjects at the Brown Institution; and on the European continent comparative pathology was poised for rapidly accelerating development in the 1870s.

CHAPTER 9

Establishing professional comparative medicine in nineteenth century France: policies and personalities

During the years of Davaine's closest involvement in the problems of anthrax, as a private individual working within the confines of a sparsely equipped private laboratory[1], there were politically motivated forces in France intent upon creating for the nation a position of leadership in the developing science of comparative medicine. From the founding of the first chair in the subject in 1862[2], to the outbreak of the Franco-Prussian War in 1870, rivalry with Germany was an effective spur to further developments; but the reasons for the many and varied French achievements in this area went deeper than that. There were a number of outstanding individuals involved. However, even native intelligence and motivation need nurturing, and that was amply provided within the French system. Political turbulence, even the revolutions of 1830 and 1848, seem never to have interrupted the steady progress of an educational system which put science, pure and applied, high on its list of priorities[3].

Its position as progenitor of the first veterinary schools had provided French science with an early advantage when simultaneous outbreaks of epidemic diseases in animals and man suggested the potential benefits to be obtained by a fusion of medical and veterinary studies. It was an approach which had received official cooperation in the work of the Academy of Medicine since its initiation in pre-revolutionary France[4]. After 1815, with improved teaching especially at advanced level in the veterinary schools, and a growing number of very active scientific societies for discussion and exchange of thought and results, there was in France a fertile environment for a budding comparative approach to medical research. It quite naturally became a joint approach for medicine and veterinary medicine as the result of the nature of the contagious diseases initially chosen as objects for study: rabies, glanders and anthrax are all transmissible to man. In the latter half of the eighteenth century, and the first half of the nineteenth century, they were at first perceived as problems for evolving veterinary medicine and legitimate objects for experimental veterinary science; gradually, as the problems of their transmission to man became increasingly obvious, medical research adopted the veterinary approach, including animal experiments, to study these and eventually other contagious and infectious diseases, until 'comparative medicine' became an established discipline in the 1840s and 1850s[5].

During the years of the greatest achievements of Rayer and of Davaine, the Paris practitioner whose often inspired research was a spare-time activity in a borrowed private laboratory, their work was complemented in France by increasingly professional and coincidentally 'comparative' experimental studies undertaken in the country's veterinary schools. The work of Chauveau at Lyons, and of Bouley at Alfort, secured for veterinary science in France unprecedented recognition and acceptance by the medical profession, which led to much fruitful cooperation.

The early career of Jean-Baptiste Auguste Chauveau (1827–1917) (Fig. 19) followed a path in complete contrast to the experiences of his medical colleagues. In his youth he had perfectly answered the notion of the ideal student as defined by Bourgelat in the early years of his veterinary schools at Lyons and Alfort[6]. The son of a farrier he had entered the school at Alfort at the age of seventeen in 1844; but with additional talent considerably beyond Bourgelat's initial requirements of reading and writing, he was able to benefit from the additional courses offered to students at this time[7]. His subsequent achievements, together with those of some of his veterinary compatriots, were clear indications of educational advantages. In a long life Chauveau became arguably the most distinguished of all graduates of the French veterinary schools. At the age of fifty a thesis on vaccinia, a subject he had by

Fig. 19. Jean-Baptiste Auguste Chauveau (1827–1917), ultimate exponent of French, and European, comparative medicine in the nineteenth century.

then made his own, earned him a doctorate in medicine[8]; and when he died in Paris at the age of ninety in 1917, replete with academic recognition and honours, his passing was mourned as that of the Grand Old Man of French comparative pathology[9].

Within the French context Chauveau was not an isolated example of the veterinarian accepted by the medical fraternity. He may have been the ultimate exponent, but he was certainly not the only veterinarian to become a member, on his merits, of the *Académie de Médecine* in Paris in the nineteenth century. His most immediate and obvious predecessor in the *Académie de Médecine* and the *Académie des Sciences*, and as Inspector General of the French veterinary schools, was Henri Bouley (1814–85)[10]. Bouley was already the second veterinarian to preside over sessions at the *Académie de Médecine*; the first had been Eloy Barthélemy, whose work on anthrax has been mentioned elsewhere[11]. Such French veterinarians could with reason afford to be proud of their profession and their background. The very existence of the office of Inspector General reflects a degree of control and consensus which lent to French veterinary education, and the profession as a whole, a strength absent across the Channel, where their counterparts were undermined by educational disadvantages and internecine administrative struggles[12]. In Paris in the 1860s, in the often heated and seemingly endless discussions of comparative aspects of various infectious diseases, veterinarians and physicians sparred on equal terms. If there was a certain amount of back-biting between them, it was no different from the sharp exchanges between physicians of opposing views[13]. Mutual respect was the order of the day.

Bouley was himself the son of a well regarded veterinarian. His elder brother became a distinguished hospital physician in Paris; Henri chose an education at Alfort, where his abilities were soon recognised. Graduating in 1836 he spent a short time in his father's practice; within a year a vacancy occurred at Alfort, and Bouley was launched on an academic career[14]. Nearly twenty years later he was elected to the *Académie de Médecine*, where he was an active member of a group of physicians and veterinarians, and a few 'pharmaceutical chemists', perhaps more correctly biological chemists, who engaged in regular weekly discussions of aspects of comparative pathology of a number of infectious diseases. Much attention was paid to anthrax and tuberculosis; recent experiments and observations of glanders formed a favourite point of reference[15]. Gaining increasing prominence was the subject of vaccinia: its origin, its identity, and its relationship to variola.

To consider this matter, and also more practical related problems of the advisability and frequency of re-vaccination (its necessity was recognised abroad well before it became accepted practice in Jenner's homeland), the Academy had established a commission on vaccinia. It was headed by Jean-Baptiste-Edouard Bousquet (1794–1872), who had for several years been in charge of the country's inoculation programmes and had written Government-sponsored reports on vaccination and on cow pox[16]. In May, 1862, Bousquet introduced to the Academy's weekly session a report from Toulouse, prepared by distinguished local physicians and veterinarians, one of whom was vaccinator at the hospital of Saint-Jacques. Based on case histories of men and of horses in the area around Toulouse, the authors' observations supported Jenner's erstwhile claim that vaccinia and cow pox had their origin in a disease of the horse. Was it, as Jenner had claimed, 'grease', or was it something more closely resembling smallpox in man and cow pox in cattle[17]? It posed a question which was to be pursued in the discussions at the Academy for several years. By 1864 Ulysse Leblanc (1796–1871), another veterinarian member of the Academy who had collaborated with Rayer in his glanders experiments referred, in a discussion on anthrax, repeatedly to 'vaccinia, or horse pox', in one place with a qualifying parenthesis of 'Ha! I know I touch here upon a burning question'[18].

Bouley made many spirited contributions to this long-drawn-out discussion. He believed the problem could be solved only by experimentation and observation, preferably carried out at his Alfort school. He had himself been convinced since 1862 that cow pox, and hence vaccine matter, had been derived, as Jenner had claimed, from a disease of the horse. Without accepting Jenner's derivation from 'grease', which had always been, at best, vague and unconvincing, the French veterinarian Petelard had in 1845 described a varioloid eruption on the face of a horse. The condition had proved to be transmissible to another horse, and accidentally to three people; to Petelard his observations suggested that 'variola' of the horse was of the same nature as vaccinia or cow pox of cattle. Bouley showed the 'vaccinogenic' disease of the horse to be a well-defined entity, and named it 'horse pox'. These discussions gain added interest in the light of recent suggestions that the origin of vaccinia virus may be a now extinct horse pox[19].

Bouley had been a student at the Alfort school at a time when belief in spontaneous generation was a creed inculcated on the minds of the students; but his open and questing mind had been swayed by Pasteur's results. His observations of the English outbreak of rinderpest of 1865–6, and the success of his own consequent quarantine measures in preventing the disease from reaching France[20], completed his conversion; he became one of Pasteur's most eloquent supporters. At the time of the discussions in the Academy, the obvious transmissibility of major epidemic and epizootic diseases was an accepted fact. Nevertheless, doubts set in in the halls of the *Académie de Médecine* when its members considered the occurrence of seemingly isolated cases of otherwise clearly contagious afflictions such as glanders and rabies, and also those which occurred in periodic outbreaks: why did measles and scarlatina, and sheep pox and foot-and-mouth disease, suddenly appear and spread when there had been no evidence of their presence since the last outbreak? Those were the kind of questions which perennially, even over the centuries, had made it easier for the layman than for the better informed professional expert to accept notions of contagiousness and transmissibility. Small wonder when we consider that even in the twentieth century difficulties remain for epidemiologists attempting to explain the behaviour patterns of that still ubiquitous scourge, influenza, to such an extent that prominent physicists can in all seriousness suggest its periodic penetration from outer space[21].

The question of the origin of vaccinia, and its relationship to horse pox, to cow pox, and to variola itself, had reverberations far and wide outside the Academy. In Lyons, the Society for Medical Science established a commission to study the problem. The initiative was Chauveau's, and he chaired the commission. Until then mainly occupied with comparative anatomy and physiology[22], he now took up the study of vaccinia and variola with enthusiasm. Under his leadership the commission carried out experiments which showed that inoculation with variola did not produce cow pox in cattle, nor horse pox in horses. Then they went further in their comparative efforts; in terms of medical ethics rather too far, as Chauveau himself was to admit after the commission's report was published. Seven children from an orphanage were inoculated with variola which had been passed through cattle and horses; of the unfortunate seven, two developed confluent smallpox, the others the discrete but characteristic form[23]. Three months later Chauveau referred at the *Académie de Médecine* to the commission's 'heavy responsibility' in carrying out the experiments, and 'in particular the experiments involving children'. There appears to have been no public outcry, and there is an almost Dickensian flavour to the episode and the implications of the commission's laconic statement that 'Transmitted to man, [variola passed through bovines] produces variola'. A generation later there was far more concern over the heroic self-experimentation of Walter Reed and his co-workers[24].

Questions of conscience apart, Chauveau's interests from then on centred on infectious diseases and their agents, in 1865 still unknown quantities. His change of direction became complete when in that year he accompanied Bouley on the above visit to London, to assess the ravages of the great outbreak of cattle plague, then at its height, and to plan preventive measures to avoid the disease reaching the French Channel ports. Bouley and Chauveau were fortunate both in the timing of their journey and in official attitudes prevailing in France. They were able to apply the lessons learnt in England to help prevent such serious outbreaks at home[25].

From then on, Chauveau was committed to the study of infectious diseases and comparative pathology, and from the 1880s also to the new immunology and the development of vaccines. In 1865 his attention at first focused on vaccinia and variola. Having abandoned the ethically suspect clinical experiments of the committee, he turned to the question of the identity of the agents of the diseases. In his first report to the Academy of Medicine on this subject, he referred to the object of his research as 'the elements which constitute the active principle of virulent vaccine matter'. Chauveau's paper was presented at the Academy by Claude Bernard, to the physiology section; the author explained that the search for putative agents responsible for the activities of the 'virulent humours' must be in the forefront of the study of the 'physiology of viruses', yet it had so far never been the subject of experimental work. It is also significant that this series of studies, in the 1860s, were published in the Academy's *Comptes rendus* in the section of '*physiologie*'; later in the century Chauveau's work was classified in the journals as '*pathologie expérimentale*'[26]. At this stage Chauveau was concerned only with vaccinia and variola, and made no reference to Davaine's anthrax studies or Bassi's earlier observations on silkworm disease. On the other hand, when Bernard presented Chauveau's second paper on the subject the following week, Pasteur was present, complimenting the author and encouraging him to widen his field of investigation, comparing his results to Davaine's anthrax studies. Pasteur was then already in a commanding position to comment. After his early chemical studies he had moved steadily towards biology with definitive papers on fermentation (1857), refutation of the theory of spontaneous generation (1862), and silkworm disease (1865)[27].

Chauveau had a clear idea of the possibilities of separating the constituents of the material he was handling. The means were filtering and decanting of the fluids; and filtration techniques were not yet adequately developed[28]. Chauveau devised a method of dilution and decantation by which he could separate out the leucocytes, leaving a suspension of particles, for which he coined the term '*granulations élémentaires*', in the serum. With added diffusion experiments he was able to show conclusively that the infective agents of vaccinia and variola, and for good measure of glanders and of sheep pox, too, were particulate and not dissolved in the serum. A similar technique was employed independently at this time by Burdon Sanderson in London, in an attempt to determine the nature of the 'virulent principle' of rinderpest. He, in his own words, set out to distinguish between a 'colloid' and a 'crystalline' agent but added: 'Diffusion was employed by M. Chauveau and myself for the same purpose; but the ideas as to the nature of the contagium which led him to adopt it were much more complete than mine'[29].

In 1870 Chauveau had at Lyons a new assistant, a recent graduate of the veterinary school, Jean-Joseph-Henri Toussaint (1847–90). From a poor background, with poor early education, Toussaint rose to acquire subsequently a doctorate in science, and one in medicine, in 1879. The same year he succeeded in cultivating the agent of chicken cholera, recently isolated by Perroncito. It was the second successful cultivation *in vitro* of a disease agent, preceded only by Koch's work on the anthrax bacillus. Toussaint supplied Pasteur with a sample of this culture; it became the basis

for the latter's work on chicken cholera which shortly afterwards
led to the development of his first vaccine[30].

Pasteur can also be seen to have benefited from the other major
study carried out by Toussaint in his all too short working life[31].
As early as 1878 Toussaint had begun a pioneering study of
microbial toxins; and following hard on the heels of Pasteur's
announcement of the possibility of vaccinating chickens against
chicken cholera Toussaint reported, in July 1880, that he had been
able to make a number of sheep resistant to anthrax infection by
means of defibrinated anthrax blood heated at 55 °C for ten
minutes[32]. Ten months later Pasteur, Chamberland and Roux
published their method of preparing an anthrax '*virus-vaccin*'
with bacilli grown at 42–43 °C, claiming that at this temperature
the bacteridium no longer produced spores, and in subsequent
cultivation became attenuated day by day. They ascribed the
attenuation to the action of atmospheric oxygen on the culture,
and also claimed that the bacteridium attenuated in this way could
no longer return to its original virulence but remained 'fixed' in its
attenuated state, a claim they were later forced to retract[33].
Scrutiny of the literature relevant to Pasteur's subsequent
demonstration of anthrax vaccination of sheep at Pouilly-le-Fort at
the end of May, 1881, reveals that on this occasion the great man
was less than candid. By then Chamberland and Roux had
improved the safety of the method of attenuation by adding small
amounts of potassium dichromate to the culture medium, a fact
suppressed by Pasteur at the time; his two collaborators were not
allowed to publish the improved method until two years later[34].
Toussaint's imperfect method was later improved by his old
teacher, Chauveau, and nor was this the last word on the matter.
It was to be further developed by Chauveau and others; and Koch
and Pasteur clashed over the possible imperfections of the method
developed by the latter. Through it all Chauveau took an impartial
view, and kept alive the interests of Toussaint, by then
incapacitated by disease[35].

Volumes have been written about Pasteur and the remarkable
development of his work, from chemistry to microbiology, animal
disease, and finally the zoonoses anthrax and rabies. Only in recent
years have French medical historians emphasised the extent to
which Pasteur, in the case of the zoonoses and the development of

vaccines, stood on the shoulders of lesser-known men. For if Toussaint (and in London, Greenfield) preceded Pasteur in attempts to develop a vaccine against anthrax, Galtier provided him with more than just useful hints for what is often perceived as the crowning achievement of his later career, the rabies post-exposure prophylactic. Like Toussaint, Pierre-Victor Galtier (1846–1908) was a veterinarian; unlike Toussaint, whose brief career had at least begun early, Galtier found his profession late and by accident[36]. Only at the age of 22 was he able to enter the veterinary school at Lyons, helped under the French system by a grant from his native *département*. Graduating top of his class, he practised for three years at Arles, and then was able to return to the Lyons school as *chef de service* in the pathology department, in July 1876. Thus the beginning of his academic career coincided with the excitement attendant on the early work in medical microbiology of Pasteur and of Robert Koch. Bouley, then Inspector General of France's veterinary schools, was instrumental in the creation in all three schools of chairs in the pathology of contagious diseases. Eminently successful in the obligatory competitive examinations, Galtier became the first incumbent of the Lyons chair in 1878 and occupied it until his death in 1908[37].

During thirty years in the chair of pathology Galtier built up both teaching and research in the discipline, studying a number of contagious and infectious diseases of domestic animals[38]. His main interests, and main contributions, were concerned with the two zoonoses, only occasionally transmitted to man, which posed very real threats in the later nineteenth century: rabies and glanders (one of Galtier's pupils, one Bonnefoy, died of glanders in 1881). He began his rabies studies in April 1879, and the first results were reported at the *Académie des Sciences* the following August. Galtier described how, after attempting to reduce the risk of experimenting with rabid dogs by developing a special restraining muzzle[39], he hit on the idea of using the rabbit as experimental animal. This was no chance development; it was a conscious decision to study in depth the disease in an animal which promised to be cheaper and easier to use than dogs and sheep. Work earlier in the century had established the fact that rabies could be transmitted to rabbits by inoculation of rabid saliva[40]; but there had been no attempt at a full and systematic investigation. Galtier set out to provide one, and in his own words, to 'dispel certain misconceptions and to fill in certain gaps in knowledge'. At the end of an extensive study he concluded:

> Rabies in the dog is transmissible to the rabbit which thus becomes a convenient and harmless subject for the determination of the virulence, or non-virulence, of various fluids from rabid animals. I have already used this type of evaluation on several occasions, in order to study the different salivas and many other fluids taken from rabid dogs, sheep and rabbits[41].

With these words Galtier established the rabbit as the experimental animal of choice for rabies research two years before Pasteur began his investigations into the disease. After describing in detail a number of inoculation experiments with rabbits, and futile attempts to prevent the disease by injection of salicylic acid, he observed that in spite of this failure the need remained to find some form of prophylactic treatment which could be applied within a day or two of exposure. Over the next two years he devoted much time to the study of rabies, its putative agent, and its effect in a number of animal species. As a veterinarian working within a veterinary school his experiments included dogs, rabbits, sheep, goats and guinea pigs[42].

In January 1881, Galtier submitted further results to the *Académie de Médecine*, including various attempts to achieve some kind of prophylactic treatment against rabies. He had found that rabid saliva injected directly into the jugular vein produced no ill effects in seven sheep, and that one of the sheep appeared to have become resistant to challenge with rabid saliva, and to injection of the same material into the peritoneum. He wrote: '[the sheep] appears to have acquired immunity'[43]. Théodoridès has analysed Galtier's work on rabies and the relative contributions of Galtier and Pasteur. Referring to a memoir by Galtier's daughter, Théodoridès pointed to the influence on Galtier of Chauveau, whose work on vaccinia and variola at Lyons had included demonstrations of the acquisition of immunity in cattle by intravenous inoculation[44]. In the same paper Galtier made a number of other important points; but he also included the inaccurate observation that the seat of rabies in dogs was exclusively in the lingual glands and in the bucco-pharyngeal mucous membrane. In future criticism Pasteur was to pay more attention to this inaccuracy than to due acknowledgement of the undoubted benefits his own studies had derived from Galtier's pioneering work.

Nevertheless, when Pasteur turned his attention to rabies in the last few weeks of 1880, Galtier had already established a factual and methodological framework of no mean order; and his first tentative report of 'acquired immunity' in sheep – one sheep – coincided with Pasteur's first note on rabies experiments, which was heard at the *Académie des Sciences* on 24 January, 1881. It was a modest beginning to what was to become perhaps the best known of all Pasteur's major works: rabbits and dogs inoculated with saliva taken post mortem from a child dead of rabies produced a disease, and showed the presence of a micro-organism, unconnected with rabies; it was a pneumococcus, and the animals died with symptoms of septicaemia[45]. Worth noting in this paper is its detailed reference to Galtier's 'esteemed experiments' of 1879, and also its conclusion: it contains the hope that if a causative agent could be found it 'might not be beyond' the capabilities of contemporary science to find a means of attenuating the virus of this 'terrifying disease'. In that case it might become possible to protect dogs, and eventually man, who is infected only 'through the carcasses or the bites of rabid dogs'. Thus Pasteur redefined the objectives laid down by Galtier two years before; from then on they continued their separate investigations into rabies and possible immunisation procedures. Galtier, the younger by more than 20 years, worked essentially alone in the veterinary school at Lyons; Pasteur, with a long established and ever growing reputation, attracted students and co-workers in considerable numbers. Above all, Pasteur was especially fortunate in some of the associates who joined him in the 1870s and 1880s. Emile Roux (1853–1933) came to Pasteur at the age of 25; he was the only one of the original group who was medically trained[46]. Charles Chamberland (1851–1908) and Pierre Emile Duclaux (1840–1904) were both, like Pasteur himself, products of the rigorous training of the *Ecole Normale*. Chamberland was a physiological chemist; Duclaux, an erstwhile student of Pasteur's, had taught chemistry at Tours and physics at Lyons before coming to Paris in 1878 and joining the newly formed Pasteur Institute in 1888. The inclusion of the veterinarian Nocard in the group completed a heterogeneous mixture of scientific disciplines and interests carrying the Institute's concerns far beyond the subject of rabies and its prevention[47].

In the initial development of a rabies vaccine, Roux and Chamberland were indispensable to the ageing Pasteur[48]. All experiments involving inoculation and injections fell to Roux, whose medical expertise was acknowledged by Pasteur, himself reluctant to undertake any such procedures. An important contribution by Chamberland had been the development of the eponymous bacteriological filter[49]; his main concern had been with water purification, after Pasteur himself had used plaster of Paris filters in anthrax studies in 1877. At the end of the century the use of improved bacteriological filters was to lead to the recognition and classification of the filterable viruses[50]. Assistance by the young Louis Thuillier was also acknowledged in the paper of January 1881; two years later a group under Roux left Paris for an ill-fated expedition to study a cholera outbreak in Alexandria, during which Thuillier died of cholera[51].

Nevertheless, on the whole the successes outweighed the failures. Three weeks after the demonstration of anthrax vaccination at Pouilly-le-Fort, Pasteur and his group could report results which had escaped Galtier. They succeeded in transmitting rabies by direct inoculation of brain tissue and cerebrospinal fluid from rabid dogs into the brains of healthy dogs. It was at last unequivocal proof of a long suspected fact: the agent of rabies was to be found not only in the saliva, but also in the brain and nervous tissue of rabid animals. The authors added a somewhat gratuitous criticism of Galtier, who had failed to transmit the disease in rabbits by inoculation of brain tissue. By direct inoculation cerebrally they were able to shorten the period of incubation to a maximum of 3 weeks. Over the next three years the group worked steadily and methodically with the one aim: to develop a vaccine against rabies.

That development and its successful outcome has been described too well and too often to need amplification here[52]. In the late summer of 1884 Pasteur felt confident enough of the efficacy of the vaccine in dogs to announce his results at the International Congress of Medicine in Copenhagen, but still expressed reservations concerning its use in man. His hand was forced and scruples overruled by compassion when young Joseph Meister, savaged by a presumably rabid dog, appeared in his office. After the recovery of Meister, the premature success of Pasteur's method of treatment caught the imagination of the public at home and abroad, at a time when rabies loomed large as a sinister if sporadic threat. Patients came to Pasteur's laboratory in search of treatment from as far afield as Russia. There were inevitable setbacks and tragedies, but there were many more apparent successes[53].

It is necessary to qualify 'success' with 'apparent'. Given the inaccuracy of diagnostic methods at the time (Negri did not describe the eponymous bodies until 1905) it is impossible to say with any certainty how many of the offending dogs were in fact rabid[54]. In recent years questions have been raised concerning Pasteur's rôle in the development of rabies vaccine. French historians of medicine have themselves been in the forefront of attempts to redistribute merit, and to point to the not inconsiderable contributions of Galtier, and later of Emile Roux. Philippe Descourt, in his partisan efforts on behalf of fellow veterinarians Toussaint and Galtier, perhaps judges Pasteur rather harshly. Théodoridès seems closer to striking a fair balance[55]. Like all history, the question of ethics in using an untried vaccine prematurely must be seen in context and against the right background: in view of the seriousness of the disease and the anxiety and despair of many patients and their relatives, insistence on more proof, more convincing scientific evidence, might have been perceived as inhuman. Nevertheless Pasteur had to contend with not inconsiderable criticism and opposition. Consequently, when the British Committee of Enquiry, appointed by the Local Government Board, reported favourably on Pasteur's method in June 1887, Pasteur himself lost no time in presenting the report at the *Académie des Sciences* at the first opportunity, in early July[56].

In spite of the setbacks, in the public mind the successes far outweighed the tragedies. On a wave of euphoria the *Institut Pasteur* was established in Paris. Pasteur himself felt that a purpose-built vaccine institute would be a suitable centre for the vaccination of patients arriving from near and far. A committee appointed by the Academy of Sciences agreed, and subscriptions were collected at home and abroad. Newspapers from Milan to Strasbourg received donations from readers; major sums arrived from the Tsar of Russia and the Emperor of Brazil, and at home from the *Chambre des députés*. The total amount of more than two and a half million *francs* handsomely covered the expenses of building the institute, with a healthy sum left over for an endowment fund. Future upkeep and running costs were left to be defrayed with the help of the sale of vaccines from the laboratory[57].

Conceived as an anti-rabies vaccination institute, the *Institut Pasteur* almost immediately became something very much more than that. Concerned also with anthrax vaccines and immunisation against diphtheria, for the group around the ageing Pasteur it developed within a few years into a busy institute of comparative pathology and microbiology, with sister institutes established abroad[58].

Before the end of the century, and in the first decade of the new century, a number of fundamental discoveries were made at the Pasteur Institute in Paris, and at its sister institutes abroad. The Paris institute was inaugurated on November 14, 1888; its *Annales* had begun publication the previous year, anticipating its completion. Roux and Yersin published their epoch-making paper on diphtheria toxin in the December issue, the first to appear after the opening of the institute two weeks earlier. Ten years later Roux, working with a number of medical and veterinary colleagues, investigated contagious pleuropneumonia in bovines; in spite of considerable merit, the resulting paper on the subject served to confuse the emerging concept of filterable viruses for many years to come[59]. As the Pasteur Institute grew and prospered, and continued to produce important work in several areas of comparative pathology, its *raison d'être*, the virus of rabies, remained an elusive entity. Its nature, its morphology, its chemical composition, continued to be unknown quantities for more than half a century, during which time the vaccine against the unknown quantity was steadily improved[60].

The success of the French institute, and its growing diversification[61], was in sharp contrast to the fate of an English institute of comparative pathology established two decades earlier.

CHAPTER 10

British comparative pathology after 1870: the Brown Animal Sanatory Institution

When Toussaint's observations were reported at the *Académie des Sciences* in July, 1880, they were certainly noted by Pasteur. It is less certain whether the latter was aware of similar developments at the same time, or even slightly earlier, in London. Chauveau may have been, and, through him, his pupil Toussaint; for as noted in the previous chapter, Chauveau was in contact with John Burdon Sanderson. Shortly after their exchange of views on the diffusion of 'virulent principles' of infectious diseases, in the late 1860s[1] Burdon Sanderson (1828–1905) became the first 'Professor Superintendent' at the Brown Institution. To no small extent did the Brown Animal Sanatory Institution owe its existence to the vision and administrative energies of Burdon Sanderson. At the time of its inception it was a unique establishment, founded with a curious bequest. 'Brown' was one Thomas Brown, Esq., M.A., LL.B., described as a 'citizen of London and Dublin' who died in 1852. Six years earlier he had made a will leaving to the University of London upwards of £20000 in investments, stipulating that it be used for 'founding, establishing, and upholding an institution for investigating, studying, and without charge beyond immediate expenses, endeavouring to cure, maladies, distempers, and injuries, any Quadripeds or Birds useful to man may be found subject to;...'[2].

Apart from the bare facts of his will, little is known of Thomas Brown. He is thought to have been a barrister-at-law; judging by the wording and provisions of his will, he was somewhat eccentric.

It is understandable that he included a time limit of nineteen years from the date of his death to the opening of the animal sanatory institution, and stipulated that if the University of London failed to meet his conditions the legacy would instead go to the University of Dublin; it is less easy to understand why in that case the funds must be used to establish there Professorships in 'any three or more' of the languages of 'Welsh, Sclavonic, Russian, Persian, Chinese, Coptic, and Sanscrit'. The testator was obviously a man of catholic interests; yet the choice of alternative subjects may not have been quite as eccentric as would appear at first sight. By mid-century Trinity College Dublin was anxious to meet demands for education of a new breed of professionals destined for the Indian Civil Service. After the loss of a final appeal for the Brown Trust the College did in fact establish a chair of Hindustani and Sanscrit a year later[3]. For not surprisingly Thomas Brown's will was contested by both next of kin and Trinity College. The Master of the Rolls eventually found in favour of the University of London, deciding that the bequest was good, useful to mankind, and a valid charity. A final appeal by the combined forces of next of kin and Trinity College was dismissed in April 1857, in spite of the fact that they were represented by the formidable Attorney-General Alexander Cockburn[4].

The opposition effectively silenced, the University of London could settle quietly to count the accruing interest on the investment; but as the end of the period specified by Mr Brown for this purpose approached, it became clear that problems remained, and that action was called for. In 1865 the Senate of the University appointed a committee with responsibility for the development of the Brown Trust[5]. A main problem for the committee was that no part of the Trust Fund could be used for the purchase of land or buildings; yet the institution with stables, laboratories, and residences for a professor and a veterinary surgeon must be situated within a mile of either Westminster or Southwark. The Senate Minute Books reflect the difficulties and the growing concern as Mr Brown's deadline approached. Two London veterinary schools offered accommodation; their offers were declined, on grounds of distance, but probably the University was less than anxious to share the Brown Trust with other institutions. The difficulty was solved only at the eleventh hour, and only by the intervention of Burdon Sanderson, who became the first Professor Superintendent. Up until 1870, Burdon Sanderson had not been involved in any negotiations, and his name appears nowhere in the Senate Minutes. Suddenly, in July 1870, there is a copy of a letter addressed to the Secretary of the Committee and signed by J. Burdon Sanderson:

> In reply to the inquiries in your letter, I am prepared to state that a sum not exceeding £4,000 will be placed in the hands of Dr. Quain and Dr. Sharpey [members of the committee, and friends of Burdon Sanderson] to be by them handed over to the London University for the purpose of establishing the Brown Institution, on the understanding that they will not apply it unless they have reason to believe that the Trust will be administered, in accordance with the Testator's Will, for the *experimental investigation* of the origin and nature of the Diseases of Animals and in so far as possible for their treatment; and that *I shall be appointed Professor with an adequate remuneration*[6].
>
> [my italics]

Initially the proposition was accepted by the Senate; five days later, the resolution was rescinded. There is no clue in the Minutes to the reasons, or if they were in any way connected with the origin of the promised £4000.

By now the deadline was looming uncomfortably near. By the autumn of 1870, the London Establishment was becoming more than a little apprehensive. The *Lancet* commented:

> The period at which the Trust Fund, if not utilised in accordance with the will of the testator, for the *study of comparative pathology* and the treatment of maladies and injuries of 'quadrupeds and birds useful to man' will be forfeited to the University of Dublin for the foundation of certain professorships in Trinity College, is drawing near. There is only, indeed, one year left to the University of London for giving effect to Mr. Brown's wishes...[7].

Already the concept of 'comparative pathology', unused by and possibly unknown to the late Mr Brown, had entered the official commentary. And Burdon Sanderson had not given up; he had been busy behind the scenes. A city merchant, one John Cunliffe, had been approached, and provided £2000 for the purchase of a suitable site. With further small sums added by Burdon Sanderson,

Sharpey and Quain, this proved finally acceptable to the University, and a property was bought on the Vauxhall to Wandsworth road[8]. At last the practical and financial problems were resolved, and the purposes of the new institution had been clarified and defined, albeit with a further twist away from the testator's original declaration. In January, 1871, the *British Medical Journal* wrote:

> Negotiations now practically completed have finally secured, for the *benefit of science and humanity*, the appropriation of this now important fund for the foundation of an *Institute of Comparative Pathology*, in which diseases of animals will be studied *in their relation to those of man*, under the charge of accomplished experts...

And the piece concluded with the writer's enthusiastic expectance of

> ... not only immediate and material benefits on the great agricultural interests of the country by the elucidation of the causes and relations of epizootic diseases, but probably greater if more remote advantages in the research after the intimate causes and origin of diseases in animals, in whom they can be *most advantageously* studied by *methods calculated to shed light on the mysteries of disease in man*[9].

It was perhaps this very readiness to embrace comparative pathology in a wider context which the following year led the Senate of the University of London to call for legal opinion concerning the study of animal diseases at the Brown[10]. The opinion obtained must have favoured the use of Brown Institution funds for such research, although it was a contentious issue which was to provide ammunition for anti-vivisectionist groups in years to come[11].

The Brown Animal Sanatory Institution finally opened in November, 1871, as a research institute with special interests in experimental comparative studies, fully equipped with laboratories and animal houses and its own animal hospital[12]. It was the first of its specialised kind, and as such a forerunner of the great institutes which were to come into existence in Paris and in Berlin within the next two decades. Unlike the Pasteur Institute and Koch's *Institut für Infektionskrankheiten*, which were created in celebration of national achievements, the emergence on the scene of nineteenth century medical science of the Brown Institution owed as much to accident as to design. France had had a chair of comparative medicine since 1862, and had a long tradition of experimental comparative medicine carried out in medical and veterinary laboratories; developments in Germany were not far behind. Britain had no such tradition; a largely demoralised veterinary profession had neither the educational background nor the spare time and energy to look beyond its own narrow horizons; and with a few exceptions, members of the medical profession rarely attempted to learn from animal diseases. Even among the papers read to the Epidemiological Society during its first thirty years, less than a handful were concerned with epizootics[13]. On the European continent, medicine and veterinary medicine had worked together for most of the nineteenth century in increasingly conscious efforts to professionalise a science of comparative medicine. There were no parallel developments in England. Yet, paradoxically, London acquired an institute devoted to the study of experimental comparative medicine before its continental neighbours. The reasons were twofold: the fortunate if accidental windfall of the Brown Bequest to London University; and the presence of Burdon Sanderson, searching for an opportunity to pursue full-time an interest in experimental pathology which had grown during his years in the Medical Department of the Privy Council and his involvement with the Cattle Plague Commission of 1865–6[14].

At its opening in 1871 the Brown Institution (Fig. 20), under the terms of the will of Thomas Brown, was affiliated to the University of London, and its Professor Superintendent had the responsibility of giving an annual series of five lectures, 'in English, and free to the public', although the University itself at the time offered no teaching and was concerned only with examinations and granting of degrees to external students[15]. Burdon Sanderson took up his new post with enthusiasm, moving his experimental work from private rented accommodation to the security of official laboratories at the Brown. The interpretation of Mr Brown's will which allowed the use of its funds for experimental pathology, and later physiology, may have been somewhat questionable; but the obligation to maintain an animal hospital on the premises was never shirked, as witness the annual reports[16].

The resident veterinary surgeon appointed by Burdon Sanderson was William Duguid. A former student of John Gamgee's at the

Fig. 20. The Brown Institution in the year of opening, 1871, from an original drawing in the Minet Library, London.

New Veterinary College in Edinburgh, Duguid had moved with Gamgee to London as a staff member at his short-lived Albert Veterinary College. At the time of the Government Inquiry into cattle plague it fell to Duguid to help with the practical details of Burdon Sanderson's research work, which was carried out at the College during and after the outbreak. Sanderson's respect for members of the veterinary profession at home and abroad reflected his admiration of Chauveau. In 1872 he wrote that he regarded himself as 'in some measure [M. Chauveau's] pupil (for there are few men from whom I have learnt more pathology)'[17]. In 1871 he did not hesitate to appoint Duguid, by now an inspector in the Veterinary Department of the Privy Council, when he himself took office at the Brown Institution[18].

Duguid's position was well defined in the statutes of the institution. Rather more vaguely outlined was a post filled when Burdon Sanderson appointed Emanuel Klein as his scientific assistant. Not yet thirty, Klein had come from Vienna with an already established reputation in histology and pathology; the authorities responded by awarding him the title of Assistant Professor. This dedicated and capable, at the same time intensely private and modest man, became Burdon Sanderson's indispensable co-worker at the Brown, where he stayed for more than twenty-five years, serving under successive Superintendents. His independent results were occasionally marred by mistakes, but he became a valued teacher and mentor to younger men, although his Viennese background continued to cause him difficulties with the finer points of the English language. His position within the University remained but vaguely defined. In 1874, when testifying before the Royal Commission on Vivisection, he was asked: 'What is your particular duty to the University [of London]?' Klein replied: 'I am ashamed to say I do not know; it has never been made clear to me'[19].

With resident veterinary help, and a very real need for fresh information in the wake of the 1865–6 disaster, Burdon Sanderson lost no time in making the study of cattle diseases, especially in epizootic form, a top priority at the Brown Institution. In 1874 the Institution joined forces with the Royal Agricultural Society to investigate two potentially serious threats to domestic cattle populations: contagious bovine pleuropneumonia and foot-and-mouth disease. The resulting study could not be described as penetrating; the investigators were hampered by a dearth of experimental animals and of suitable cases of the natural disease in the neighbourhood, combined with legislative difficulties due to restrictions on the movement of infected animals. Reports on inoculation experiments published between 1876 and 1879, with complementary epidemiological reports by Duguid, make it abundantly clear that pleuropneumonia, foot-and-mouth disease, and anthrax were all present in frequent outbreaks in the British Isles during these years, as they had been on and off for centuries[20].

Burdon Sanderson made no attempt to pretend that his studies on pleuropneumonia were other than incomplete. In his concluding paper he regretted that his 'so far satisfactory' experiments with regard to the use of 'inoculation as a means of preventing [the spread of pleuropneumonia]' had had to be curtailed and in fact 'brought to an abrupt termination' because of 'legislative difficulties'. What results he had obtained remained inconclusive because of the small number of animals available for experimentation[21]. There can be little doubt that the legislative difficulties were not confined to questions of movement of infected animals; they included the difficulties which the Brown Institution had to fight throughout the crucial years of its pioneering work: the frequently misinformed, sometimes violent, and mostly unreasoning opposition of the anti-vivisectionists. It was a peculiarly British problem; its much milder manifestations on the European continent even when they did occur, as experienced for example by Claude Bernard and later by Pasteur during his rabies work, were often inspired and orchestrated by British movements. On the occasion of the Norwich vivisection trial in 1874 the *Lancet* wrote:

> The French medical press have freely commented on this trial, and the writers maintain that the British Society for Preventing Cruelty to Animals has needlessly gone out of its way to display a maudlin sympathy where none was needed. Dr. Magnan had performed the same experiments at Lyons at the time of the Congress; and they proved highly interesting, the scientific aim held in view being approved by all lookers-on. The French writers point out with much emphasis that a sister society exists in France, but that it has contrived to escape the quicksands of over-sensitiveness....[22].

The imbalance in obstacles presented to animal experimentation in the two countries may have contributed to French superiority in the fields of comparative physiology and pathology in the nineteenth century. A more serious defect in the British system was the absence of veterinary research practised by the veterinary profession. The legacy of inferior education at London's Royal Veterinary College bestowed by the long years under Coleman had created a body of veterinary surgeons ill equipped for research activities; and the alarming inadequacy of defence against the cattle plague outbreak of 1865–6 had convinced the majority of the veterinary profession of the futility of academic study of germ theory and related subjects. Experience had taught them that in island communities only strict adherence to time-honoured methods of animal epidemiology could offer hope of success in coping with outbreaks of infectious animal diseases. Emphasis must remain on quarantine, isolation of infected herds, and the veterinarians' privilege in epidemic situations: slaughter[23]. The rise of veterinary science on a scale when it could even begin to compete with its French and German counterparts occurred only in the last decade of the nineteenth century and was largely the work of one man, John McFadyean[24]. Meanwhile the veterinary assistants at the Brown Institution remained, especially after the departure of Duguid in 1877, shadowy figures who, although contributing to the research undertaken by their Superintendents and a number of visiting investigators, were given little or no credit in print. In his history of the Institution Sir Graham Wilson spared only half a page to the names and work of four of the ten known veterinary assistants between 1871 and 1939, plus a brief mention of the ignominious case of A. M. Porteus, who vanished without trace, having absconded with a sum of more than £100, which had to be written off as a bad debt[25].

In addition to contagious bovine pleuropneumonia, foot-and-mouth disease had been mentioned in a preliminary report on experiments made at the Brown Institution during its early years, but it made no appearance in Burdon Sanderson's later reports. This may have been due in part to the fact that the threat of outbreaks had temporarily receded, as witnessed by Duguid's annual reports on the health of farm animals[26]. More importantly, by the time Burdon Sanderson left the Institution, in 1878, focus at the Brown had shifted to anthrax studies, an English counterpart to the studies of Davaine, Chauveau, Toussaint, and Pasteur in France, and Koch in Germany. From being a disease mainly observed 'in the so-called anthrax districts on the Continent' it was, in the 1870s, making alarming inroads among animals on English farms. It was an obvious and pressing problem for study at the Brown Institution. Burdon Sanderson, assisted by Duguid, launched an inquiry into aspects of the disease. The results were published only two years later, as an introduction to a first report by his successor at the Brown, William Smith Greenfield[27].

In experiments carried out between February and June, 1878, his last year at the Brown, Burdon Sanderson had established two important facts. One was the transmissibility of anthrax 'poison' through foodstuffs, which remains an important path of transmission, especially in colder climates. Of the other he wrote:

> When the disease is transmitted by inoculation from cattle to small rodents, such as guinea pigs, and then from them back to cattle, the character of the disease...is much milder than that of the original disease acquired in the ordinary way. The rodents die, but the bovine animals inoculated with their blood or with the pulp of their diseased spleens recover[28].

This was the promising basis inherited by W. S. Greenfield when he took over as Professor Superintendent at the Brown in December, 1879. With the limited resources at his disposal he made the most of it.

For Burdon Sanderson his departure from the Brown marked the end of his formal involvement in experimental pathology, and his activities from then on were commensurate with his duties first as Professor of Physiology at University College, London, and ultimately, from 1895 until his death ten years later, as Regius Professor of Medicine at Oxford. But he remained true to his early interests and the subject of his early, most productive years. When in his will he left a bequest to the University of Oxford it was specifically for 'the support of the *pathological department* of the university and especially to provide for the expenses of research in *pathology* in the said laboratory or elsewhere'[29].

The first seven years of the Brown Institution's existence, under Burdon Sanderson, were important ones. True to the spirit of the testator, the main object of research in the institute's laboratories and in the field had been cattle disease; and important links had been forged between the institution and the Royal Agricultural Society, to their mutual benefit. The Brown provided expert scientific guidance; the Society in turn supplemented the meagre financial resources of the Brown for key investigations. When Greenfield took over, a certain amount of scepticism was voiced regarding his suitability to lead 'The one place which the Government had been able to look to for really original research in great questions of pathological research and comparative hygiene'[30]. Greenfield soon confounded his critics. In the Brown Lectures for 1880 and 1881 he disclosed not only his considerable insight into the rapidly evolving science of medical and veterinary bacteriology, but also his successful development of Burdon Sanderson's and Duguid's unfinished search for acquired resistance to anthrax. In this work he anticipated Pasteur and even Toussaint, if only by a few months. It was a parallel development in France and in England, reflecting the intensity of the search at this time for effective measures against the ubiquitous threat of anthrax. None of the three early methods proposed by Greenfield, by Toussaint, and by Pasteur, Chamberland and Roux, was perfect.

The methods were all based on attempts to attenuate the agent. Greenfield's method consisted simply in successive cultivations[31]. Toussaint heated anthrax blood for a short time at 55 °C[32]. Pasteur and his co-workers originally heated their 'bacteridium' culture in air at 42–43 °C and claimed that this prevented the formation of spores and produced a 'fixed' attenuated agent permanently unable to revert to full virulence. As seen above, the latter statement had to be retracted and the method improved by the addition of potassium dichromate to the culture medium. Greenfield and Toussaint carried out their experiments independently and virtually simultaneously, between December 1879 and early 1880. They published their first results within a few weeks of each other, Greenfield on June 5 and Toussaint on July 12, 1880. Pasteur, Chamberland and Roux published their first paper on the subject in 1881, and the dramatic demonstration at Pouilly-le-Fort took place in late May of that year. The amended procedure did not appear until 1883[33]. Pasteur's announcement of his fortuitously discovered attenuation of the chicken cholera agent (*Pasteurella multocida*) was made at the *Académie des Sciences* in November, 1880. On this occasion the *Lancet* pointed out that the most important aspect of Pasteur's communication had 'been already anticipated by one of our own investigators in the same field. The fact is that successive cultivations may attenuate a virus which, thus weakened, still confers protection against the inoculation of the virus in the most virulent form'[34].

The following year, after the demonstration at Pouilly-le-Fort, and when Greenfield had moved to the University of Edinburgh as Professor of Pathology, he summed up his feelings in his inaugural address in the following way:

> M. Pasteur has recently published the results of a very large series of experiments made by a precisely similar method, and with results fully confirming those which I published more than a year ago. And although I venture to claim for England whatever merit may be due to priority for the discovery, I none the less rejoice that the facts should have been so fully established in France. My experiments were made with a small and inadequate sum of money furnished by the generosity of a private society, and in the face of all the difficulties interposed by law; whilst M. Pasteur is encouraged and abundantly supplied with means by the liberality of the French Government[35].

The 'private society' was the Royal Agricultural Society. The dispassionate tone of Greenfield's dignified statement made no attempt to hide his feelings of frustration in the face of underfunding and Government neglect. His sentiments were echoed in the same year by John Simon when he criticised the Cruelty to Animals Act at the International Medical Congress in London. Referring to Greenfield's inoculation experiments Simon, no stranger to the capriciousness of Government intervention, quoted the former's bitter reaction to the necessity of becoming a licence-holder under the Act; and also his conviction that the Act presented such obstacles to the investigator that it defeated all attempts to help animals at risk, and discouraged legitimate and potentially valuable studies[36].

Burdon Sanderson and Greenfield were the two Superintendents whose work at the Brown was concentrated in the field of comparative pathology. Under the next two professors, experimental physiology increasingly made its presence felt. In 1871, on the eve of the opening of the Brown Institution, the *Lancet* had been at pains to point out that '...physiological investigations form no part of the scheme of the Brown Institution, which will be devoted entirely to pathology'[37]. Now things were to change rapidly. Before 1870, distinctions between pathology and physiology had been only vaguely drawn. Pathology was frequently referred to as 'pathological physiology'. As late as 1873, the *Lancet*, reporting on an 'Address in Physiology' to the British Medical Association by Burdon Sanderson, made little distinction between the two disciplines and their laboratories, commenting that 'The establishment of the Brown Institution is one of the most important scientific events that have ever happened in this country; and the approaching opening of the splendid physiological laboratories of Owens College, under Dr. Arthur Gamgee, will form a second and very considerable advance in the path of progress'[38]. The change in emphasis towards experimental physiology was to culminate later in the century, when according to Stephen Paget the Brown was of great importance, not only as a veterinary hospital, but as a recognised centre in London for advanced research in pathology and physiology[39].

Greenfield's immediate successor at the Brown was Charles Smart Roy (1854–97), who stayed little more than two years. Earlier he had spent some time at the Brown studying the pathological anatomy of contagious bovine pleuropneumonia; he had also worked in German physiological and pathological laboratories. As Superintendent he engaged in an extensive study of the physiology of the mammalian heart, interrupted only by a journey to Argentina to investigate a serious outbreak of cattle disease. The disease was anthrax, and for Roy it was an opportunity to test the so recent results of experiments on attenuation and protective inoculation obtained by his predecessors at the Brown and, more publicly and spectacularly, by Pasteur. Two years after the demonstration at Pouilly-le-Fort Roy found, in experiments in the field in the Argentine Republic, the flaws in Pasteur's published method. In his Brown Institution Report for the year 1883 he explained his attempts to 'mitigate the virus' of anthrax by passage in the local prairie dog. He found the attenuated material suitable for protective vaccination of horses and cattle. When trying to confirm the 'assertions of Pasteur' concerning the protection of sheep in a similar way he had problems. He wrote: 'My observations showed, however, that certain precautions are necessary in carrying out this method, and that without these failure is almost certain'[40]. It was Roy's last

Brown Report; in 1884 he was appointed to the newly established Chair of Pathology at Cambridge, where he remained until his early death in 1897[41].

Roy was followed at the Brown by Victor Horsley[42]. The young Horsley was already committed to neurophysiology. By one of the fortunate accidents which can, and sometimes do, shape the history of institutions, the 1880s when he took over at the Brown coincided with a change of emphasis in contemporary comparative pathology which accorded well with his preoccupations. With the successful development of a vaccine against anthrax, its problems had become less immediate; at the same time concern over the threat of rabies was growing in towns and cities in France and in Britain. In April, 1886, the Local Government Board established a commission to evaluate Pasteur's post-exposure prophylaxis and Horsley, not yet thirty, became its Secretary. Public concern was justified: with outbreaks among foxhounds and stray dogs there were inevitably human fatalities. The worst affected areas in Britain were Lancashire, the West Riding of Yorkshire, and London. Hardly a week went by without the *Lancet* reporting some disquieting incident; a particularly unhappy one concerned five boys from Poplar, bitten by a rabid dog on their way home from school in July, 1885. None survived. George Fleming, Inspecting Veterinary Surgeon at the War Office and author of *Rabies and Hydrophobia*, wrote an alarmed letter to the *Times*. In December, 1885, the *Lancet* commented:

> We have skilled experimenters, familiar with the region of work over which PASTEUR's labours have ranged. We have a Brown Institution, in which material abounds, and among its legitimate objects this may surely take the foremost place.[43]

The most severe of zoonoses, and caused by a neurotropic virus, rabies presented an apt focus for the research of Victor Horsley, neurophysiologist, Superintendent of the Brown Animal Sanatory Institution, and Secretary to the Rabies Commission. With other members of the Commission he went to Paris that same April, to see at first hand and to evaluate the method practised in Pasteur's laboratory. Initially doubtful, they were soon convinced of its importance. Pasteur, anxious for recognition abroad, cooperated with the Commission and provided Horsley with two rabbits to take home to form the basis for experiments. When eventually the

Commission's report was published, in June 1887, it confirmed Pasteur's results and offered such unqualified support for his method of prophylaxis that he himself presented it at a meeting of the *Académie des Sciences* in early July. Pasteur was in particular need of support at this time[44]. That the support was mutual was acknowledged by Horsley years later, when he was giving evidence before a Royal Commission on Vivisection. On that occasion Horsley drew on the experiences of the Rabies Commission in Paris in 1887, referring to Pasteur's views and to the animal experiments which had resulted in the successful development of a vaccine against rabies[45]. Horsley's faith in Pasteur's method remained unshaken after he lost a laboratory assistant, bitten by a rabid cat, who died in spite of being sent to Pasteur for treatment. The case was unfortunate in the extreme, but not difficult to explain. The patient, so Horsley pointed out, did not count sobriety among his virtues; and while under treatment he repeatedly escaped to spend his nights in sordid drinking bouts. During one such he had to be rescued after falling into the Seine. In the circumstances, Horsley was less than surprised at the fatal outcome[46].

By the time Horsley came to give evidence before the Royal Commission, the British Isles had been free from rabies for more than five years. The successful eradication of rabies (it returned briefly following World War I between 1918 and 1922, after which Britain has remained free from the disease) was due to a campaign spearheaded by Horsley. It involved the introduction, and more importantly the enforcement, of the Muzzling Order, and of strict and lengthy quarantine regulations still in force today. Horsley's campaign received unstinting support from Walter Long (later Lord Long of Wraxhall, 1854–1924), an astute politician who in 1886 became Parliamentary Secretary to the Local Government Board, and in 1900 its President. The hard but ultimately successful fight fought by Horsley and Long in the teeth of hardened opposition from the anti-vivisectionists and their allies, the Dog Owners' Protection Association, is reflected in the pages of the *Lancet* and of the *British Medical Journal* for these years[47].

The cooperation between Horsley and Long was a model of its kind, a milestone in public health achieved by ideal collaboration between medical and veterinary science (originating at the Brown) on the one hand, and parliamentary politics and legislation on the other. Advances in medical understanding in the late nineteenth century had greatly improved its chances of success; a success based on Pasteur's advice to Horsley in Paris many years before, when he had remarked: 'Why do you come here to study my method?... You do not require it in England at all. I have proved that this is an infectious disease: all you have to do is to establish a brief quarantine covering the incubation period, muzzle all your dogs at the present moment, and in a few years you will be free'[48].

During his years at the Brown, Horsley paid attention in equal measure to the institute's twin research subjects of experimental pathology and physiology. His work on rabies was carried out at the same time as fundamental research into the function of the thyroid gland, and studies in neurophysiology and neurosurgery[49]. His successor at the Brown Institution was Charles Sherrington (1857–1952). Sherrington had studied the new pathology and bacteriology in Germany in the laboratories of Virchow and of Koch; once at the Brown he, too, turned to neurophysiology, although, like Horsley, he found the obligations of the institute's rabies testing service a help rather than a hindrance in his research[50]. He left for a chair in physiology at Liverpool in 1895, in what was becoming almost an established pattern for Professor Superintendents at the Brown. Under its next two superintendents, John Rose Bradford (1863–1935) and Thomas G. Brodie (1866–1916), the emphasis remained on physiology[51].

The focus of research reverted to pathogenic organisms under the Brown Institution's last superintendent, F. W. Twort. He accepted the appointment in 1909, and remained until the buildings in the Wandsworth Road were destroyed by enemy action in 1944. His later years were marred by controversy and unhappy disputes with the authorities[52]. At a time when identification and cultivation of pathogenic organisms, and the failure to grow 'filterable viruses' in artificial culture, occupied the attention of a majority of medical bacteriologists, Twort's approach was unusual; he was rewarded with unusual results. He had long been interested in the effects of different conditions of growth on bacteria, and believed variations in growth conditions to provide a possible key to isolation of pathogenic organisms such as the bacilli of tuberculosis and of leprosy[53]. Once at the Brown, he turned to the bacillus of paratuberculosis in cattle, the so-called Johne's disease. Primarily a disease of horned animals, not transmissible to man, it was an eminently suitable subject for research at the Brown. It was also work which Twort carried out in close collaboration with his veterinary colleague at the Brown, G. L. Y. Ingram; and Ingram received full credit as a co-author[54]. But for Ingram's early death in 1914, veterinary science could at last have had its due at the Brown Institution.

Twort's early interest in the effects of different conditions for growth led him to develop ideas which were to provide the framework for his later, single-minded work: the concept of pathogenic organisms being derived from 'wild' types which already existed in nature. Long before present preoccupations with ecology, Twort believed that different environmental conditions could account for differences between pathogenic organisms and non-pathogenic ancestors. He hoped that if he could find non-pathogenic forms of filterable viruses, they might prove easier to cultivate than the pathogenic types. It was this vain hope which brought him to work on vaccinia, and to the accidental observation of the phenomenon with which he is eponymously linked: the 'Twort phenomenon', or the lysis of bacteria by bacterial viruses, later known as bacteriophage[55]. In the 1930s, and more especially after World War II, the study of bacteriophage, in Europe and in America, has led to their use as protagonists in the explosive development of twentieth century molecular biology[56].

Twort's work of lasting importance, and his fundamental discovery of the bacterial viruses, were made during his first ten years at the Brown Institution, immediately before, and during, World War I. At the end of the war, Twort's increasingly anguished appeals to the University for improvements to his own and the Institution's position fell on deaf ears. His own research failed to meet its pre-war standards, and the Institution went into a terminal decline long before its site and buildings were finally destroyed by enemy action in 1944[57]. Its accelerating decline in the twentieth century was hardly surprising. From the turn of the century it had been losing its unique status at home as well as abroad. New institutes had been emerging in the wake of the creation of the Pasteur Institute in Paris, built to produce rabies vaccine and funded initially by public subscription; it soon received additional government support and widened its scope to include other areas of comparative pathology[58]. So did the *Institut für Infektionskrankheiten* in Berlin, created for Robert Koch as the German answer to the Pasteur Institute. Pasteur Institutes *outre-mer* followed in French possessions abroad, and other vaccine institutes proliferated in many countries and other continents, all soon to pursue also the study of comparative pathology on a wider basis[59].

Ultimately it was not only competition from abroad which sealed the fate of the Brown Institution. In London, the Lister Institute was established just before the turn of the century. It was better funded, better administered, better equipped, and it had a *raison d'être*, as well as a source of income, from the production of vaccines, diphtheria antitoxin, etc. It also offered instruction in public health[60]. It represented a kind of competition which the Brown had no means of challenging. At the same time, veterinary education and veterinary services in the capital, and in the country as a whole, were rapidly improving, and consequently the numbers of animal patients brought to the hospital at the Brown decreased steadily, from a peak of over 7000 in the first decade of the twentieth century, to an all-time low of 1000 on the eve of World War II[61]. In the circumstances it was only surprising that the Brown Animal Sanatory Institution managed to survive, in the form it took, for as long as it did.

CHAPTER 11

Nineteenth century developments in comparative medicine on the European continent

The patterns of gradual evolution of comparative medicine in the nineteenth century were discernibly different in the major European countries. Reduced to its essentials, the discipline emerged when animal medicine combined with human medicine to provide a foundation, and a necessary framework, for a germ theory; a theory which could begin to exist in earnest only with the benefit of animal experiments based on specific infectious diseases attacking both man and one or more other mammalian species. It was the relative contributions of medicine and of veterinary medicine, and the background and approach of their practitioners, which differed from country to country.

Some reasons for divergence have been noted in previous chapters. In France, home of formal veterinary education, there was from an early stage, and especially after 1815, exemplary collaboration between the medical and veterinary professions, whose members combined peaceful coexistence and cooperation with mutual respect. In Britain on the other hand, until quite late in the century, cooperation remained largely out of the question because of the particular circumstances surrounding its main veterinary school. The reactionary attitudes of its principal and its governors fed old prejudices, and kept the veterinary profession locked in internal conflict and ensuing external impotence. In the circumstances there was minimal cooperation between the two professions; veterinarians, with no training or encouragement for scientific research, concentrated their efforts on management and control of animal diseases with scant regard for the implications of zoonoses, except occasionally in the case of rabies. Where medical scientists worked on comparative aspects of infectious diseases, as during the early years at the Brown Institution, their relations with veterinary colleagues were those of master and assistant; there was never any doubt who presented the results, and who received the major credit.

The Italian and German states, with their decentralised administrations for the better part of the century, presented yet other, and different, patterns of evolution. In Italy, the eighteenth century had seen the once fiercely independent republics, duchies and city states accept foreign ascendancy from Spain, Austria, and France; until the Napoleonic wars political apathy prevailed. The scientific community on the other hand had lost none of the enthusiasm of earlier centuries, and the establishment of the French veterinary schools in the 1760s elicited immediate response in medical faculties in major Italian universities. In Italy as elsewhere in Europe the continued threat of cattle epizootics, with which the medical profession had had to deal earlier in the century, added urgency to the desire for trained veterinarians; although as noted above Bourgelat's schools had initially shown a preference for the equine species and its diseases for which authorities in other countries were unprepared.

Francesco Bonsi[1] (Fig. 21) was an Italian contemporary and, in a sense, counterpart to Bourgelat. Like Bourgelat he was a sprig of an aristocratic family, educated for the law, whose abiding interest in natural science and horsemanship turned him in the direction of veterinary medicine. Although playing some rôle in the creation of veterinary schools in Rome and in Naples, Bonsi, within the decentralised Italian system, never took on the country-wide practical authority of Bourgelat. His greatest contributions were in his writings, covering a span of more than half a century from 1751 to shortly before his death in January, 1803. His early work was concerned with the horse and horsemanship; but from 1782, when he returned to his native area of Rimini, he turned his attention to the cattle epizootics still periodically invading Italy from Dalmatia, and along the Adriatic coast. His observations on cattle plague, first published at Rimini in 1786, were reprinted at Florence after the Tuscan outbreaks of 1796 and 1800, at the time when Michele Buniva was recording the Piedmontese epizootics[2].

Several Italian States had been quick to take advantage of the

Fig. 21. Francesco Bonsi (1722–1803) and eighteenth century veterinary art (courtesy Wellcome Institute Library, London).

French veterinary initiative and to send students to the French schools, although again, the decentralised nature of political administration militated against a concerted national effort. Individual departments had their own local preoccupations; but their paths towards comparative approaches may well have been eased by relative isolation and the proximity within that isolation of local medical schools, from which some members of their staff had been recruited in the first place. By the end of the eighteenth century, veterinary education was well established in a number of Italian universities.

Shortly after the opening of the Lyons school, Turin, capital of Piedmont (then part of the Sardinian kingdom of the Dukes of Savoy), had sent four students from its medical faculty to follow the courses there. Among the four was Carlo Giovanni Brugnone (1741–1818), who stayed in France for a total of five years, and who on his return to Turin became the director of its newly established veterinary school, the first in Italy[3]. The geographic and political divisions in the country may even have introduced an element of competition which resulted in the establishment, in rapid succession, of no less than seven veterinary schools and departments in Italy before 1800. In all these cases the initiative came from the medical faculties. This might not have pleased Bourgelat; his discouraging response to Cicognini's polite inquiries in 1772 has been recorded above[4]. Nevertheless, students of medicine and surgery continued to arrive at Lyons and Paris from abroad, and they were not turned away. Many of them were instrumental in creating veterinary schools or departments in their respective countries and universities upon their return[5]. Cicognini's Milan finally acquired a veterinary school in 1791. At the time, Lombardy's Austrian overlords would accept only a compromise solution in the form of a one-year farrier course teaching also minor surgical intervention. In such circumstances, most of the staff responsible for educating veterinarians in Italy before 1800, however sound their background, and however extensive had been their own earlier education in France, were able to offer little beyond the teaching of farriery and comparative anatomy. Only with the decrees imposed by Napoleon after the installation of Eugène Beauharnais as Viceroy of Italy in 1805, did the country acquire veterinary schools with full courses modelled on the French ones[6].

By then Italy, and Piedmont in particular, had lived through other serious outbreaks of rinderpest in the 1790s. If the chief protagonist of the ultimately successful efforts to control the epizootics was, as had been the case with epizootics at the beginning of the century, a physician, he now had the advantage of close links with some of the men trained in the French veterinary schools. M. F. Buniva[7] was a friend and colleague of Brugnone. He had French contacts of his own, and an abiding interest in smallpox and in Jenner's work on vaccination. All these factors combined to give him a broad perspective reflected in his writings on the epidemiology of rinderpest in his native Piedmont[8].

Buniva's reputation as promoter of Jenner's vaccination method was matched in Milan by that of Luigi Sacco[9]. In 1809 Sacco wrote a treatise on vaccination, in which he included chapters on grease and on pox disease in sheep[10]. He attempted to clear up some of the uncertainties surrounding the nosology of Jenner's 'grease', no easy matter as the disease was unknown in Italy at the time. In fact its exact relationship to vaccinia has never been fully established, although according to Baxby some experiments quoted by Sacco could have intriguing implications in retrospect[11]. In his chapter on sheep pox, however, Sacco had the confidence of first-hand experience, and his accompanying illustrations (Fig. 22) can take their place alongside those of J. Beart Simonds, which were published forty years later (chapter 6). Sacco carried out a number of independent experiments with vaccinia, proving its prophylactic qualities. He made a point of thanking the enlightened authorities for their support and for 'placing at [his] disposal orphans for public experiments'. He appears to have had no scruples concerning the ethics of his use of children from the Santa Caterina Foundling Hospital for his experiments, any more than had the Princess of Wales about inoculation experiments on prisoners in the previous century[12]; nor is there any evidence that he felt a need to apologise later, as Chauveau was to do in 1865[13].

There is only circumstantial evidence to link Sacco's interest in and knowledge of sheep pox to his proximity, geographically and intellectually, to Giovanni Pozzi[14] (Fig. 23). They were both erstwhile students of Johann Peter Frank (1745–1821), at Pavia and at Vienna; they both settled in Milan, where Sacco became chief physician to the *Spedale maggiore*, and where Pozzi was charged, from 1807, with the organisation of the full veterinary school when it was finally being raised to French standards[15]. Pozzi taught at the school and wrote a textbook of pharmaceutical chemistry as

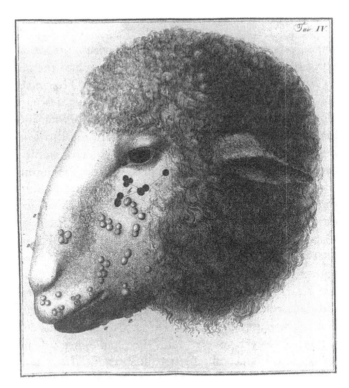

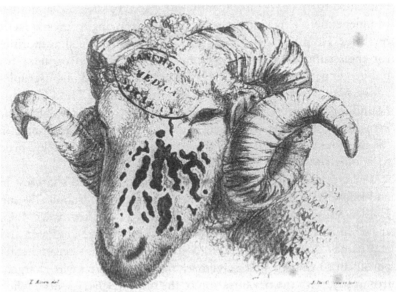

Fig. 22. Illustrations of sheep pox by Sacco (1809) and by J. B. Simonds (1848) (courtesy Wellcome Institute Library, London).

Fig. 23. Giovanni Pozzi (1769–1839) (courtesy Wellcome Institute Library, London).

applicable to man and animals. The title pages of each of its two volumes carry Pozzi's maxim: 'Medicine of man, and medicine of beasts, are sisters'[16] – a worthy sentiment for someone with his training and background, and one which Coleman in London would certainly have disputed. Pozzi's works also included a treatise on movement in animals and man, in which he declared his allegiance to John Brown and to the Brunonian theory. His main work on disease was a labour of love and loyalty to his old teacher when, after the death of J. P. Frank, he completed his *Complete System of Medical Polity* with a volume on leprosy[17].

A most important, most basic, and ultimately most seminal Italian contribution to the development of comparative medicine came later, two decades after Napoleon's final defeat, and from a most unexpected corner. Agostino Bassi[18] was essentially an amateur scientist: a civil servant with a degree in law, who had also followed some of the courses in natural science at Pavia. Brought up in the country, as an adult he spent his spare time exploring methods of improving local agriculture. He wrote extensively on various breeds of sheep, before his attention was caught by a disease of silkworms threatening the Italian silk industry. Known in English as muscardine, one Italian name for the disease reflects the existence of chalky-white patches on the bodies of dead and dying silkworms: *calcinaccio*. By painstaking microscopic work (which contributed to the eventual loss of his eyesight) Bassi was able to establish the cause of muscardine: a parasitic fungus, transmitted from worm to worm and growing in their bodies to produce the characteristic appearance. As his eyesight continued to deteriorate after the publication of this study in 1835[19], Bassi turned increasingly to speculation on analogies with other diseases. In an age when theories of the causation of infection and contagion were still very much a matter of contention, he confidently claimed that cholera, too, was caused by 'a germ of a cryptogam, mould, or fungus', and that smallpox, spotted fever, bubonic plague, and syphilis, were also produced by 'living parasites, vegetable or animal'. Bassi's analysis of the pathology of muscardine was the first correct identification of a microscopic disease agent based on scientific observation. It was also responsible for the tendency in early bacteriology to identify most disease 'germs' as microscopic fungi[20].

In the German States, the approach to veterinary education, animal disease, and comparative medicine also developed to a pattern of its own. In the early German veterinary schools there was far more administrative uniformity than in Italy. This was largely because Prussia had emerged during the later eighteenth century as the dominant German state, the richest, the largest, and politically the best organised. Most of the early German veterinary schools were placed within its widely scattered provinces. They were meticulously planned and, as in Italy, medical students and graduates with appropriate abilities and interests were encouraged by state bursaries to supplement their education by visits to veterinary schools abroad, and later at home.

From their early days, Bourgelat's schools had attracted the interest of Frederick the Great, acutely aware of the losses inflicted on his people and his armies by diseases of horses and of cattle. His personal physician recommended the creation of a veterinary school already in the late 1760s; building plans for a Berlin school were enthusiastically supported by the King and by the Academy of Sciences; but the country's – and the King's – costly wars left no funds available for such ambitious projects, and the school was still at the planning stage when Frederick died in 1786[21]. In the event, the first German veterinary school to survive was established at Hanover in 1778. The authorising decree was signed by George III, who pointed to the pressing need for well educated veterinarians 'in Our German possessions' where especially the royal studs and cavalry regiments required trained 'horse doctors and farriers'[22]. There is a curious contrast between the king's enthusiasm for a school in his Hanoverian residence and his apparent indifference to the fate of the London infirmary and school planned, but abandoned for lack of support, by Edward Snape in the same year of 1778[23]. It may to some extent have been due to George III's sentimental regard, in later life, for his German roots. On the other hand, he never visited his duchy of Hanover, and the initiative is unlikely to have been his own. Hanover's affairs were in the hands of a regency composed of a few of its leading families. The king's handwritten authorisation, referring by name to the 'Oberhoftierarzt' Kersting, who became the school's first director, and his young assistant and eventual successor, Havemann, was signed at St James's Palace on July 7th,

1778. In Hanover, no time was lost in issuing the printed decree, rewritten by the Hanoverian authorities, on July 18th, 1778. Clearly the Hanoverian regency, and Kersting, who had already begun preliminary moves to arrange teaching, wielded a great deal more power and influence than Edward Snape and his few supporters in London[24].

The French Revolution and its repercussions did little to favour development in veterinary science in Europe until the end of the Napoleonic Wars. The emphasis was on training, as rapidly as possible, as many veterinarians as possible, to tend the animals of the armies and cavalry regiments. On the other hand, in a far corner of the German States, in the middle of the wars, a physician in Jena made a contribution, which is often cited as a milestone in rabies research. In fact, it was more than that. It was a first attempt to perform controlled animal transmission experiments with a contagious disease. G. G. Zinke's observations have been discussed in detail elsewhere[25].

After Napoleon's final defeat, in 1815, Europe settled into a period of relative calm and preoccupation with peacetime concerns. The incipient interest in animal disease and, through it, in comparative medicine, which had begun in pre-revolutionary France, was re-kindled and was soon a major object of research in European centres. It was not surprising that attention should focus initially on two diseases, both zoonoses, which emerged in their different ways as threats to man and his animals throughout Europe at this time. The one, rabies, had made its presence felt through the centuries; it was now increasingly becoming a major problem in expanding towns and cities, where growing numbers of dogs were being kept as pets or, worse, as fighting dogs. The other one was glanders. This had long been recognised as a destructive presence in populations of that most valued of domesticated animals, the horse. The economic importance of glanders had been particularly striking during the many wars fought on European soil in the past[26]. Now, with more emphasis on exact clinical observation, and a growing interest in animal experimentation in the early nineteenth century, glanders emerged as a focus of attention. There was a sudden accumulation of observations pointing to a fact gone unnoticed, and certainly unrecorded, before 1800: glanders was shown to be a zoonosis, a disease although primarily of horses, transmissible on occasion to man; like rabies, with fatal results.

The fact that these two much studied diseases are primarily diseases of animals, underlines certain basic circumstances often overlooked in evaluations of animal experimentation. In the field of pathology, unlike in those of anatomy and physiology, animal experiments were not initially designed solely as models for the study of human disease. They first came to prominence in attempts to investigate disease processes in animals of economic importance, in order to cure or prevent their own diseases. Because a number of the diseases studied in the nineteenth century, above all anthrax, were zoonoses, it gradually came to be realised that such studies could serve as models for the elucidation of disease processes in man.

The early transmission experiments with rabies and glanders have been discussed elsewhere; so have some of the later ones with anthrax[27]. It may be emphasised here that the architects of these studies all relied on veterinary connections, perhaps with the exception of Zinke; but then he was almost certainly influenced by the work of John Hunter and his associates[28]. K. H. Hertwig, on the other hand, was a typical exponent of the Prussian system, which encouraged and even subsidised those interested in combining medical and veterinary studies.

After an initial two and a half years of medical studies in Breslau, Hertwig was given government grants for further studies in veterinary medicine and medicine at other German schools. Five years later, having completed a veterinary degree, he obtained an appointment at the Berlin Veterinary School, where he taught without neglecting his medical studies. Another five years later he graduated as doctor of medicine. From his work on rabies[29], to his later teaching and his writings on diseases of dogs and of horses, and his official journeys as far afield as Russia to recommend preventive measures against importation of rinderpest, he amply repaid the cost of his subsidised studies[30]. His exact contemporary E. F. Gurlt[31] was another successful representative of the German system of subsidies, which functioned in spite of wars, difficulties and disease in the early nineteenth century. Both Hertwig and Gurlt rose above their difficult beginnings to positions of authority in the main Veterinary School in Berlin. Together they founded the

Magazin für die gesammte Thierheilkunde in 1835; for forty years they remained joint editors.

One of the more colourful characters engaged in early anthrax research had also benefited from a dual education under the German system, although geographically by a different path. Born in Weimar, F. A. Brauell[32] studied medicine and veterinary medicine in Jena, Berlin and Copenhagen. Born in the first decade of the nineteenth century, he belonged to a generation of German scientists whose talents were sought after by Russian universities. After working in Vilna and Kazan, he joined the new Veterinary Institute in Dorpat at its inauguration in 1848[33]; and it was here that he found anthrax bacilli in the blood of his veterinary assistant at post mortem examination, as in the blood of three experimental animals which the deceased had been helping to inoculate[34]. Combined medical and veterinary interests informed all the early work on anthrax. Among Brauell's contemporaries, Barthélemy's background was in veterinary medicine, as was that of Delafond. Rayer, only peripherally involved in recognising the *bactéridie* had, if no veterinary degree, strong connections with the profession, and Davaine was his student[35].

It was on this bedrock of medical and veterinary achievement that Robert Koch came to build when, as a young general practitioner in the depths of the German countryside, he used his spare moments to erase the uncertainties left by Davaine's work on the disease which also affected a few of Koch's patients, and considerable numbers of their livestock. It was with Koch's further determined work, and his perfection of techniques, that the science of medical and veterinary bacteriology came into its own. Where Pasteur had been guided at each stage of his career by the need to solve specific problems, from the diseases plaguing the wine industry, and the maladies of silkworms, to the immunisations of sheep against anthrax and dogs against rabies, Koch's approach and motivation were different. They represented the born researcher's single-minded search for the unequivocal evidence of pure science, and for the methodology which could produce such evidence[36].

Koch and Pasteur were rivals in science, each protecting an international reputation; the upheavals of the Franco-Prussian War exacerbated a situation which must always be difficult. Closer to home, Koch had less than easy relations with an equally eminent adversary. When Robert Koch entered university at Göttingen in 1862, Rudolf Virchow, at the age of forty, was one of Germany's, and Europe's, best known and most admired physicians[37]. He had edited the great *Handbook of Pathology and Therapy* which appeared in 1856, to which he himself contributed chapters on diseases caused by plant parasites and by animal parasites, and on zoonoses: glanders, rabies, and anthrax[38]. His 'Cellular Pathology' had appeared in 1858; and ever since his visit to Upper Silesia in 1848, the year of revolution in Berlin, Virchow's political consciousness and concomitant determination to work for improvements in hygiene and nutrition for the underprivileged, had been growing. Also, he became increasingly convinced that cooperation between medicine and veterinary medicine would be necessary in the fight to achieve control of infectious diseases. His broadly based views, his ability to include a diversity of factors in his considerations, made him react coolly to Koch's narrower preoccupations with isolation of causative agents[39].

Virchow's use of dogs as experimental models in his major work on thrombosis and embolism has been recorded in an earlier chapter[40]. His concern for public health and his interest in veterinary science alike are reflected in his many publications, over a period of twenty-five years, on trichinosis in man and animals, and the life-cycle of its parasite[41]. Bovine tuberculosis was another interest, and one in which he did not see eye to eye with Robert Koch. Well before the work on anthrax by Koch, by Galtier, and by Pasteur, Virchow had shown that blood from deer suffering from anthrax caused death when injected into the veins of rabbits, whereas sheep were not affected by such inoculations[42]. Nor did he forget animal disease in his political life. A founder member of the German progressive party, and always active in its opposition to Bismarck, he also championed agricultural interests. In 1877 he sponsored, in the Prussian Chamber of Deputies, a resolution urging the government to enforce strict international undertakings to control the spread of rinderpest[43].

Koch's rise to fame, nationally and internationally, was swift and well deserved. Over a period of five years he moved from the publication of his first paper on anthrax to the dramatic announcement of his discovery of the tubercle bacillus[44]. From then on, identification of specific organisms pathogenic in animals and man continued in rapid succession until the end of the century. It was a development made possible by Koch's perfection of the pure culture method, and of techniques for staining and photomicrography of such cultures[45]. It was a great achievement, but in none of its aspects was it an entirely new departure. Koch learned from a number of predecessors; it was his inspired synthesis of such received knowledge, and its further development, which earned him his place in the history of microbiology.

Pasteur had grown bacteria in liquid media, attempting to 'purify' the cultures by serial transfer. Koch's mentor, Ferdinand Cohn, had studied non-pathogenic bacteria, and his pupil, Joseph Schroeter, had obtained cultures on slices of potato, and on an enriched starch paste[46]. Photomicrography had been performed with various objects, of biological and non-biological origin, since shortly after the invention of photography, and the improvement of microscopes, in the 1840s[47]. Staining methods had been pioneered by the pathologists and histologists; in Germany especially by Carl Weigert and his cousin Paul Ehrlich[48], in the United States at the Army Medical Museum in Washington, DC, by J. J. Woodward in the 1860s and early 1870s. Woodward's report on the pathology of bovine pleuropneumonia dated June, 1870, often overlooked[49], contains perfect examples of early photomicrography[50]. The specimens used by Woodward were supplied by John Gamgee, who was acting as consultant to the US Commissioner of Agriculture on pleuropneumonia and other epizootic diseases during his early years in the United States[51].

Brock's recent balanced account of Robert Koch's life and career demonstrates the difference in quality between the early pioneering work on pure cultures and its corollary, the triumphs of the discoveries of the tubercle bacillus and the cholera vibrio, and later less impressive episodes[52]. After that, both the world, and Koch himself, came to believe in his infallibility, with unfortunate results. His premature and highly publicised claims for the therapeutic value of tuberculin could only dent his reputation in the long run. In the short run, it led to the creation of the *Institut für Infektionskrankheiten*, as a research institute for Koch, with a clinical department devoted to tuberculin treatment[53]. It was from this institute that Koch set out on his voyages when, from his position of established eminence, he was urged to help combat African epizootics, as well as malaria and sleeping sickness. His late work in Africa, like his less than candid behaviour during the tuberculin episode, lacked the inspiration and the painstaking accuracy which had ensured the monumental and lasting importance of the work on anthrax, on pure cultures and solid media, and on photomicrography, the basis for the eponymous postulates and for the identification of bacteria[54].

Apart from the anthrax studies, Koch's work on animal disease belongs to his Africa years, and the results were less impressive than the early work. In the 1890s, outbreaks of rinderpest had been slowly spreading in Africa, from north to south. In 1896 the disease finally crossed the Zambesi River and progressed through the Transvaal and the Orange Free State to threaten the Cape Colony[55]. The authorities there, acting independently of London[56], appealed to Robert Koch for help to solve the problem, and Koch arrived at the Cape on December 1st, accompanied by a high-ranking German army physician. As always, Koch began in a spirit of pure science, by looking for a causative organism. Unable to find, let alone cultivate, agents of the disease in infected blood[57], he resorted to attempts to develop methods of immunisation. For all his earlier criticism of Pasteur, his proposed procedure owed a very great deal to his rival's work. Carefully omitting reference to Pasteur by name, he wrote magisterially: '...we have many instructive analogies. Thus we know the microbe of neither small-pox nor of rabies, and yet we have succeeded in devising prophylactic inoculations against both diseases dependent upon the fact that the infective material can be weakened and converted into a so-called vaccine...'[58]. Even so, Koch's results were less convincing than had been those of Pasteur. The circumstances were of course very different: different animals, different specificity, different epidemiological patterns. It is a measure of the general belief in the possibilities of the new scientific approach, created by the work of Pasteur and of Koch in the last quarter of the nineteenth century, that the authorities so readily put their faith in Koch and immunisation procedures, rather than rely on the slaughter and quarantine policies which had held sway since the early eighteenth century[59].

Less than four months after his arrival in Cape Town, Koch was ordered by his own government to leave for India, to head the German Plague Commission in Bombay. Protesting his reluctance to leave the rinderpest research unfinished, he reported to the *British Medical Journal* his belief that he had 'solved the main problem in its essential points'[60]. In fact, much was left for others to do, and there was to be no easy solution to the question of a vaccine against rinderpest. Initially, Koch's unsatisfactory method was replaced by one developed by Arnold Theiler[61].

Born in Switzerland, where he had graduated in veterinary medicine, Theiler had settled in Johannesburg in 1891. Combining a keen mind with practical ability, he was placed in charge of the campaign to control rinderpest in Transvaal in the crisis year of 1896. Unlike Koch's half-hearted attempts, Theiler's approach to epizootics and their control was that of a dedicated professional veterinarian; his work was to be of lasting importance throughout the difficult years of the Boer War and World War I. His early interest was in African horse sickness; he had found its agent to be a filterable virus when McFadyean published similar results in London in 1900[62]. Long after Koch's brief and over-publicised visits to Africa, Theiler continued to improve his methods. Working with other veterinarians, among them Stewart Stockman[63], his native talent in the long run outshone Koch's modest achievements in Africa. Towards the end of World War I he was still working on African horse sickness and a method of immunisation based on principles similar to those employed in immunisation against rinderpest: simultaneous inoculation with immune serum and live virus[64].

Rinderpest was not to be Koch's only encounter with animal disease in Africa. In July, 1897, he returned from India, this time to German East Africa, to be based in Dar es Salaam, where there was also an outbreak of bubonic plague. It turned out to be less serious than at first assumed, and Koch could turn his attention to other diseases present in the area. They included malaria, surra[65], and other diseases of cattle of greater economic importance, which Koch confidently pronounced to be in 'symptoms, blood manifestations, and post mortem findings' so similar to the well known, and at the time in the United States much studied, Texas fever, that he did not hesitate to identify it as such[66]. For good measure, he included illustrations of the 'parasites of Texas fever (Pyrosoma)'.

Cranefield in a recent study has shown that in this case Koch's report was 'wrong in almost every possible way' and, furthermore, misleading to later students of the disease, which can now be identified as East Coast fever[67].

In view of the somewhat meagre results of Koch's involvement with epizootic disease in Africa in the 1890s, it may seem surprising that in 1902 his help was again requested, this time in Rhodesia. Concern had been growing over 'Rhodesian red water', which was in fact identical with East Coast fever, both caused by the tick-borne parasite *Theileria parva*, named for Theiler who in the end did a great deal more for control of the prevailing epizootics in Africa than did Koch, and to rather less public acclaim[68]. Dwork has documented the struggles preceding the invitation to Koch between, on the one hand, Joseph Chamberlain and the British Colonial Office, who wanted a competent British representative to deal with the Rhodesian crisis, and the authorities in Rhodesia, who preferred Koch and his still dazzling reputation in spite of the relative failures of his earlier visit to the African continent[69].

The lack of any convincingly successful results which characterised Koch's work with animal diseases in Africa, was not reflected in studies carried out in his absence in his institute in Berlin. Cattle disease of a different kind was threatening German livestock. As Koch had done initially when studying rinderpest in Africa, his associates Loeffler and Frosch at home, on behalf of a worried German Government Commission, were attempting to isolate an agent responsible for the present foot-and-mouth disease, which 'every year does millions' worth of damage to the country's agriculture'[70]. Already in their preliminary report, published in September 1897, Loeffler and Frosch could establish that all bacteria so far isolated by others, and tentatively identified as causative organisms in foot-and-mouth disease, were merely accidental contaminants. They singled out for comment the so-called Siegel–Bussenius bacillus isolated earlier in Berlin[71]; they forbore to mention the 'micrococcus' of which Klein in 1886 had claimed that he had ascertained that it was the true cause of the disease[72].

In their full report six months later, the German investigators had been able to draw far more startling conclusions: the agents of the disease were sufficiently minute to pass easily through bacteria-proof filters; but they also were still able to produce the disease after repeated dilution, and so must be able to multiply. With Beijerinck's simultaneous work on tobacco mosaic disease, the results marked the beginnings of a new discipline, which was to evolve slowly in the twentieth century: that of the filterable viruses[73].

When in his later years Koch so willingly and enthusiastically undertook the rigours and discomforts of travel in Africa in the pursuit of diseases of animals and man, he came up against diseases which could not immediately be explored by the methods so successfully devised for bacterial infections. The agent of rinderpest, like that of foot-and-mouth disease, belonged to the filterable viruses, and could not be grown on the artificial media developed for bacteriological research. Those of malaria, of surra (*Trypanosoma equinum*, *T. evansi*), and of East Coast fever (*Theileria parva*) were parasites with complex life cycles. The study of parasitology had begun elsewhere, and was already in competent hands by the time Koch arrived in Africa[74]. By 1900, crucial work on malaria had been done by Laveran in French Algeria, by Manson and Ross in British possessions, and by Bignami and Grassi, and by Celli, in Italy; Griffith Evans, veterinary officer in the Punjab, had transmitted surra to a healthy horse with blood containing trypanosomes in 1880, and by the turn of the century, the scene was set for the long-drawn-out saga of Bruce and Castellani, of sleeping sickness, tsetse-flies and their trypanosomes[75]. Also in Italy, the first chair of parasitology was created in 1875 for Perroncito, the veterinarian who pioneered the study of the transmission of animal parasites to man with food, and who went on to identify hookworm as the cause of severe, and sometimes fatal, anaemia among those working on the St Gotthard tunnel[76].

By the end of the nineteenth century, increasing political stability had allowed the initially different patterns of development of veterinary education and medical and veterinary collaboration in the main European countries to settle into similar individual frameworks within which comparative medicine could flourish. Although the Pasteur Institute always retained its original function as the main anti-rabies vaccine establishment, it also has continued to grow over the years to its present position of eminence in all matters concerning infectious diseases[77]. The *Institut für Infektions-krankheiten* could not for long maintain the illusion of offering a cure for tuberculosis by means of tuberculin, which instead became a valuable diagnostic tool. The study of foot-and-mouth disease was an early example of the versatility of the staff of the institute. Although all the people brought in by Koch had had a medical education, Paul Frosch did, in the same year as undertaking the investigation of foot-and-mouth disease, begin teaching bacteriology at the Berlin veterinary school[78].

The two great pioneering institutes were soon followed by many lesser ones, whose achievements were far from negligible. Most of them began as vaccine institutes, with emphasis on rabies treatment, and developed to offer other services, as well as doing basic research. The Brown Institution, perennially short of funds, and with no possibilities for diversification, was inevitably left behind in the competition at home and abroad. Developments in the twentieth century were to take place elsewhere.

CHAPTER 12

From European nucleus to world-wide growth of institutes of comparative medicine

When Robert Koch embarked on his seven years of successive expeditions to Africa and India in 1896, he had behind him the memory of triumph in Egypt in the previous decade. The discovery there of the cholera vibrio in 1883[1] belonged to the happy years of his early successes, rivalling the isolation and cultivation of the tubercle bacillus, and without association with the later disappointments of tuberculin. The fact that the simultaneous mission from the Pasteur Institute failed to find the agent of cholera, and ended with the tragic death of one of their number, Thuillier, from the disease[2], may not have unduly disturbed Koch.

The cholera expeditions were representative of emerging attitudes to public health at government level in Europe in the late nineteenth century. Throughout the century European colonialism had been growing; and the economic importance of maintaining the health of indigenous working populations, as well as of Europeans overseas, was becoming increasingly obvious. Already in the eighteenth century, the health of both civilian seamen and troops abroad had been a cause for concern, as documented in a number of reports, whose subjects were most frequently scurvy and the perennial puzzles of tertian and quartan fevers[3]. By the middle of the nineteenth century, new improved microscopes came to the aid of the investigators. Before the advent of the new bacteriology, with its full complement of staining and pure culture techniques, and Koch's postulates, protozoa, larger and hence easier to identify microscopically, were being incriminated as causative organisms in a number of what became known as tropical diseases[4]. Parasitology was evolving side by side with bacteriology, and the need for effective means of putting the new knowledge into practice was everywhere felt. At the same time the demand for rabies prophylaxis was increasing; and the treatment of diphtheria with anti-diphtheria serum was developing rapidly following the discovery of toxins and anti-toxins[5]. There were political as well as humanitarian reasons for building more institutes, and training more staff, to meet the needs both in Europe and overseas.

From the earliest days of anti-rabies vaccine, demands for institutes established along lines similar to the Pasteur Institute in Paris had been expressed from near and far. Countries where rabies presented immediate problems, and which were geographically too distant to allow patients easy access to treatment in Paris, were among the first to establish their own centres for preparation and distribution of vaccine. The first to be built outside France was in Russia; and in South America, Brazil was the first to acquire an institute, in the northern coastal town of Pernambuco (now Recife). Many more followed elsewhere in short succession. All these institutes owed their existence to the need for rabies vaccine; all of them became institutes of bacteriology and comparative pathology within a short span of time[6].

In Britain, the Rabies Commission had evaluated and approved Pasteur's method by the summer of 1887; but for the time being, the authorities were content to enforce the Muzzling Order and quarantine measures. Patients bitten by rabid dogs were sent to Paris for treatment; from 1885 to 1889 a total of more than 200 unfortunate victims of dog bites had to undertake the uncomfortable journey across the Channel[7]. In the summer of 1889, the realisation that such numbers of British patients had to rely on the charity of what was essentially a free service at the Institut Pasteur, finally spurred politicians and the medical profession to action. At a meeting at Mansion House, the Lord Mayor of London heard evidence from leading authorities on rabies, including Victor Horsley, Burdon Sanderson, and James Paget; letters of support were received from the Prince of Wales, T. H. Huxley, and John Tyndall[8]. As a result, a fund was established, with the dual purpose of supporting the Paris institute with a donation, and of meeting future expenses of sending patients to Paris. Also among the meeting's resolutions was a request for a Government Bill to introduce simultaneous muzzling of all dogs throughout the British Isles as well as extensive quarantine regulations for all imported dogs[9].

It was the committee formed to collect and administer the Mansion House Fund which, after completing its immediate task, proceeded to discuss the possibility of establishing in London an institute with aims similar to those of the Institut Pasteur in Paris. It was to be an institute for research, as much as vaccine production; at a time when the antivivisectionists were particularly active, this was a sensitive issue, but not the only obstacle to realisation of plans for the British Institute of Preventive Medicine, legally incorporated in July, 1891[10]. True to British traditions in the nineteenth century, the foundation of the institute was envisaged as dependent on private subscription rather than public funds provided by the Government; and the problems of finding and purchasing suitable land and buildings were as acute, if not as protracted, as had been those of the Brown Institution twenty years earlier.

At the time of incorporation of the institute, its aims stated in the Memorandum of Association were remarkably similar to those laid down for the Brown Institution, except for additional commitments, most importantly to provide 'instruction and education in Preventive Medicine to Medical Officers of Health, Medical Practitioners, Veterinary Surgeons and advanced Students'[11]. There was no specific mention of rabies and hydrophobia, and reliance on prophylactic treatment in Paris continued until the disease had been eradicated in Britain in 1902. The production of vaccines was, however, an integral and important part of the Institute's work throughout its history, and its successive changes of name, from the BIPM, to the Jenner Institute, to the Lister Institute; but the main vaccines it supplied were against diphtheria, bubonic plague, and smallpox[12]. The institute's commitment to education was emphasised by its amalgamation, in 1893, with the College of State Medicine, formed in 1886 to providing training for Medical Officers of Health[13]. Ten years later, as the Lister Institute for Preventive Medicine, it acquired the status of a recognised school within the University of London[14].

In spite of similar aims, and often similar difficulties, there appears never to have been any suggestion of cooperation, let alone merging, of the Brown Institution and the Lister Institute. The years of the Brown's decline between the wars were years of discovery and scientific achievement at the Lister, where the much better facilities allowed simultaneous research into a number of diseases of animals and man. Important work on cattle diseases included studies on pleuropneumonia and on foot-and-mouth disease, investigated with such enthusiasm at the Brown in the 1870s and 1880s. Half a century later, at the Lister, better facilities and improved methodology yielded results of a different order. Patterns of pleuropneumonia infection in laboratory rats and mice gave clues to the interpretation of the epidemiology of the disease in cattle. The work on foot-and-mouth disease was, to minimise the risk of spread of infection, carried out on an obsolete warship moored off Harwich, under the auspices of the Ministry of Agriculture[15]. The team involved in the investigation was led by Joseph Arkwright, and even in the early stages, in the floating laboratory, results were obtained which had implications for the control of the disease: it could be transmitted to guinea-pigs and a number of other small mammals, which could be used in experimental infection. This line of enquiry led to the realisation that the disease is transmissible within natural populations of hedgehogs, which may thus act as carriers over considerable distances[16].

The Lister Institute became a school of the University of London in the same year as the London School of Tropical Medicine. In the following year, the two institutions found themselves on opposite sides in a controversy concerning a chair of protozoology endowed by the Colonial Office[17]. E. A. Minchin, appointed to the chair, chose to go to the Lister, which offered more space and better facilities. Until his early death ten years later, when the Institute's department of protozoology disappeared, Minchin and his associates contributed to the study of life cycles of trypanosomes, in animals in the laboratory, and those of sleeping sickness in Africa[18].

The institute had been involved in attempts to control disease abroad since its early days as the BIPM. In 1896, before plans for the London School of Tropical Medicine got under way, Patrick

Manson, then physician to the Seamen's Hospital at Albert Dock[19] was treating there a sailor who had returned ill from the East. Biopsy of an affected lymph gland, examined at the BIPM at Manson's request, showed the presence of the recently discovered plague bacillus[20]. It was the institute's first professional encounter with a disease which was to figure prominently in its reports, and in its production of vaccines, for many years. At first it was a domestic problem, sparked off by a small epidemic in Glasgow in 1900, and it was viewed as such by the Local Government Board under Walter Long – the same Long who with Victor Horsley had carved out the rabies legislation. It soon became evident that the production of Yersin's anti-plague serum, and Haffkine's fluid, was relevant not just as a means of building up stock for possible future emergencies; supplies were urgently needed by the Colonial Office for distribution abroad. The outbreak in India which had brought a number of Plague Commissions from several European countries to Bombay during the later 1890s[21] showed no signs of abating. It was a constant threat to far-eastern countries and colonies.

The plague bacillus had been known since 1894[22]; the spread of the disease in man following the routes of migrating infected rats was at last a recognised fact; and its transmission by fleas had been suspected, and shown to be likely, by Simond and others[23]; but there was no conclusive evidence, and no means of control. By the autumn of 1904, with no improvement in the situation, the India Office, with the help of the Lister Institute, moved. An Advisory Committee was formed, funds were contributed by the Indian Government, and a carefully selected research team left to carry out field work in Bombay in the spring of 1905[24]. After three years of painstaking research, they had impressive evidence for the transmission of bubonic plague from rat to man by flea bites. This extensive investigation also included detailed studies of the species of flea involved and, an important factor in India, the degree and durability of infection present in different types of cow-dung flooring contaminated with the plague bacillus[25].

With plague research in India, and sleeping sickness investigations in Africa, the Lister Institute joined an international effort to control transmissible diseases in what has since become known as the Third World. Apart from such individual contributions as

Koch's late travels and, from Britain, expeditions from the Lister Institute and from the London School of Tropical Medicine, campaigns to study and eventually to control diseases of animals and man in unpropitious climates have often centred on purpose-built laboratories created in strategic outposts. Among the first and most successful were the Pasteur Institutes *outre-mer*, institutes of microbiology and comparative pathology after the pattern of the Paris institute as it developed and expanded from its original purpose of combating rabies. The first was established by Albert Calmette at Saigon, at the beginning of 1891[26]; four years later, a second was founded by Yersin at Nha Trang, further up the coast from Annam (South Vietnam)[27].

The young Calmette had come to the Paris Institute in 1889, to follow Roux's course in microbiology, after serving as a physician in the French navy. Roux soon recognised his qualities, and less than two years later, he was on his way to Saigon. It is a measure of his success there that at the end of the decade he was back in France, to direct a new institute at Lille, the second within France itself. The Saigon institute's priority was the control of the epidemic diseases prevalent in Indo-China, by the production and distribution of vaccines against smallpox and rabies, and by developing prophylactic measures against cholera and dysentery. But there were also other problems tempting the enquiring mind of someone trained by Roux and Pasteur. Snake bites were another hazard of life in the East, and Calmette pursued the study of their toxins and antitoxins, and the development of a serum therapy[28].

The Pasteur Institute at Lille was planned with different aims in mind. The need which brought it into existence was for anti-diphtheria serum, treatment for a dreaded disease made possible by the recent discoveries of exotoxins, and of anti-toxins[29]. Experimental work on smallpox vaccine, which at Saigon had included the use of young buffaloes for vaccine production, was continued at Lille; and the problems of hookworm infections in the mining communities of northern France also occupied Calmette during those years, as did the production of anti-plague serum[30]. Of all his varied activities, it was his work, with Guérin, on tuberculosis, and the eponymous vaccine BCG (Bacille Calmette–Guérin) which stands out among his achievements. Long before its conclusion his search for effective preventive measures, for education in basic hygiene and outpatient care, led to the foundation at Lille, as early as 1891, of a clinic offering advice and treatment to those not accommodated in hospitals and sanatoria. It was an example of Calmette's determination to serve the community in a social as well as a medical sense, and the '*Dispensaire Emile Roux*' became a model for similar establishments elsewhere[31].

Although he helped to establish a Pasteur institute in Algiers in 1910, Calmette did not again work abroad after his return from Saigon, to recover from dysentery, in 1893. The rest of his career centred on the institute at Lille and, after World War I (when he had been interned during the German occupation of Lille), the Paris institute. The career of Yersin, his colleague in the Colonial Medical Corps, who founded the Nha Trang institute in 1895, followed a different path. After Calmette's return to France Yersin directed both institutes, known collectively as the *Instituts Pasteur d'Indochine*; and he never returned to work outside the Far East. His published work had begun with his collaboration with Roux on bacterial toxins; in 1894 in Hong Kong he isolated the plague bacillus (*Yersinia pestis*)[32]. Later, the work of the Nha Trang institute reflected Yersin's taste for comparative studies in a wider sense even than the majority of Pasteur institutes, traditionally pursuing the many aspects of Pasteur's own varied interests[33].

Throughout the nineteenth century, developing medical science, especially in microbiology and comparative pathology, had been generally regarded as a European field of endeavour. The specialist laboratories, and the many outstanding scientists who worked in them, were visited by aspiring American students as a matter of course. Towards the end of the century, a number of those American students had returned to influential positions within the educational system of the United States, and its medical schools. Emphasis on research was introduced on equal terms with clinical teaching[34]. At the same time, private endowment for education, and especially medical education, was introduced in the United States on a scale unknown in Europe. If the situation was never completely reversed, the exchange of students and scholars between Europe and the United States in the twentieth century came to work both ways; and whichever way the exchange, it was financed by American institutions more often than not[35]. In the medical sciences, the main source of philanthropy, at home and abroad in the twentieth century, has been the Rockefeller family; it is less than surprising that the first, and ultimately the most influential, institute of medical science and comparative pathology in America should bear their name.

John D. Rockefeller had founded the University of Chicago in 1893; during the later 1890s, Mr Rockefeller's philanthropic plans were guided more specifically towards medical science by his adviser, F. T. Gates[36]. At the end of the century, Gates had sown the seeds of his own growing interest in medical philanthropy in the mind of his employer. Plans for a medical research institute with Rockefeller endowment were beginning to take shape. The great European institutes had been established to consolidate and develop great medical discoveries and resulting therapy; the Rockefeller Institute for Medical Research, with no such pre-determined point of departure, perceived its purpose in more general terms. The institute's sphere of interest would be in 'the sciences and arts of hygiene, medicine and surgery…in the nature and causes of disease and the methods of its prevention and treatment,…'[37]. Although Gates had made a determined effort to inform himself of the latest developments in medical research and practice, and in turn to enlighten Mr Rockefeller, they needed professional help to shape their plans. The initial Rockefeller contribution of $200 000 was intended as an enabling grant to explore the possibilities of successfully establishing an institute of the kind envisaged by Gates and Rockefeller. In the absence of an existing framework, such as had already been in place in the cases of the *Institut Pasteur* and the *Institut für Infektionskrankheiten*, the first step was the formation of a Board of Scientific Directors whose responsibility would be the realisation of the planned institute.

Ideas of scientific medical research linked to clinical teaching within established medical schools had been gaining ground in the United States in the last decade of the nineteenth century. By the year 1900, a number of the more progressive medical schools in the better known universities had acquired research laboratories, where a new generation of medical scientists, most of them with some training in Europe behind them, were eager to establish their subject as an integral part of the medical curriculum[38]. The most comprehensively planned and equipped department opened as the

Johns Hopkins Medical School in 1893, with William H. Welch as professor of pathology, beginning a life-long association which culminated in the directorship of the University's School of Hygiene[39]. His position at Johns Hopkins, as well as his scientific achievements, made him a natural choice for inclusion on the Board charged with organising the Rockefeller Institute.

Welch's early studies in Europe had been carried out in Cohnheim's laboratory, and his work on wound infections and diphtheria toxins established impeccable credentials in comparative pathology[40]. Even more indicative of the way in which research interests at the new institute were to extend and widen the horizons of existing research laboratories was the presence of his colleague on the Board, Theobald Smith, since 1895 professor of comparative pathology at Harvard University[41]. Ironically Smith, whose work epitomised the comparative approach, and who more than any of his contemporaries in the United States made fundamental contributions to the study of animal diseases, had had no chance to study in Europe in his youth. Coming from a poor immigrant background, he made his way by state scholarships through university and medical school. Only after graduation did he realise his bent for pure research. In 1885, he began his career as director of the pathology laboratory of the Washington based Bureau of Animal Industry. It was in this capacity that he carried out his best known and most seminal work, the study of Texas fever. Also known as Southern cattle fever, the disease had long troubled, and puzzled, farmers and stock-owners by its persistence and its curious patterns of distribution and transmission[42].

A decade earlier, Patrick Manson, working under primitive conditions in China, in the town of Amoy, had discovered the path of transmission of blood parasites of the genus *Filaria*, and their development in an intermediate host, the mosquito[43]. Now Theobald Smith, working with the veterinarian F. L. Kilborne, carried out an extensive investigation of Texas fever, demonstrating in painstaking detail the complex aetiology of the disease: the transmission of the protozoan *Babesia bigemina* through a development stage in blood-sucking cattle ticks[44]. Their report is a classic monograph containing the first definitive proof of spread of an infectious disease by an arthropod vector. Definitive proof had escaped Manson, who was working on *Filaria* infestation in man; Smith and Kilborne had the advantage of not just the resources of the United States Bureau of Animal Industry, but also the opportunity to make extensive animal experiments. It was a model study of its kind. During the decade following its publication in 1893, other tropical diseases were shown to be vector-borne. Ronald Ross, inspired by Patrick Manson, demonstrated the mosquito's rôle in the transmission of malaria[45]; Reed and Carroll, in what can only be described as heroic experiments, did the same for yellow fever and its filterable virus[46].

Another vector-borne protozoal disease of economic importance to American animal industry was blackhead of turkeys and chickens. Theobald Smith began his study of this disease during his early years in Washington, and only completed it more than twenty years later at the Rockefeller Institute[47]. His first paper on the subject was published in the year of his move to Boston, when in 1895 he was appointed to a new chair of comparative pathology at Harvard University. It was from this position, and his concomitant seat on the Board of Health, that he published a comparative study of human and bovine tubercle bacilli. In his conclusion he suggested possible ways of resolving '...the want of unanimity as regards the channels of infection' between the views of Koch and of others. The way forward must lie in 'The study of the bacilli in primary intestinal disease in children in which the source of infection is assumed to be outside of the family and possibly in the milk'[48].

Such were the respective backgrounds of William H. Welch, the outgoing cosmopolitan with a talent for administration, and Theobald Smith, the introverted home-grown scholar, when with five other well known medical scientists they joined the Board responsible for planning the Rockefeller Institute, newly certified by the State of New York[49]. A first step was the appointment of a director, a position initially offered to Theobald Smith, who felt himself unable to leave Harvard so soon; instead the Board settled for Welch's young protégé, Simon Flexner. He was to lead the institute for more than thirty of its crucial early years[50].

The opening of the Rockefeller Institute in 1904 coincided with a period of rapid development in the medical sciences it had been created to serve. Bacteriology was by now a well established discipline, its roots firmly in the great European institutes; its recent off-shoot, the discipline concerned with the filterable, or ultravisible, viruses, was the subject of speculation and experimentation in laboratories, and in the field, around the world. The recent successes of teams and individuals from the U.S. Army Medical Services studying the aetiology, transmission, and epidemiology of yellow fever in Cuba and the Panama canal zone[51] emphasised the importance of the avowed objectives of the new institute. It was one problem among many in need of the attentions of trained investigators suitably equipped. By 1904 the proper facilities for the Institute were in place; and within its first decade, the Rockefeller Institute attracted an impressive staff of young and promising researchers. Among the results obtained, some were merely competent; others stimulating and controversial by their very unexpectedness. It facilitates the historian's task that from a very early stage in its existence, the Institute acquired that emblem of 'having arrived' a journal of its own. It happened less by design than by unforeseen circumstances.

In the early years of the Johns Hopkins Medical School, W. H. Welch had perceived the need for a specialist journal serving the new breed of American medical researchers working within that institution. In January, 1896 was launched the first issue of *The Journal of Experimental Medicine*, with Welch himself as editor, cooperating with a 'board of associate editors who are leading representatives of their special departments in the United States and Canada'[52]. But Welch did not have the temperament to make an effective editor; and by the turn of the century the new journal was in trouble. Five volumes had been published; volume 6 languished in a half-finished state while efforts were being made to persuade others to relieve Welch of his self-imposed task which he now found too arduous. The Rockefeller Institute finally consented; and volume 7 appeared in 1905, with Simon Flexner and E. L. Opie as editors, and prefaced by the terse statement that 'The *Journal of Experimental Medicine* will hereafter be published under the auspices of the Rockefeller Institute for Medical Research', and assuring prospective readers that the change of management would not affect the contents of the journal[53].

Simon Flexner had been well prepared for the editorship as well as for the directorship of the Rockefeller Institute by his association with Welch. Trained in experimental pathology in the pathological laboratory organised at Johns Hopkins by Welch, within the medical school which Welch as Dean did so much to shape[54], he brought knowledge of administration and enthusiasm for research to the task in hand. The news in 1908 of the successful transmission of poliomyelitis to Old World monkeys by Landsteiner and Popper[55] was immediately taken up by Flexner; with P. A. Lewis he showed the agent to be a filterable virus in the same year as Landsteiner's own demonstration[56]. Soon afterwards Landsteiner left the field of virus research for immunology, and it was as an immunologist that he joined the Rockefeller Institute after World War I. Flexner's group continued work on poliomyelitis, and other virus diseases, for many years. It was not the only example of Flexner's flair for promising aspects of other people's research. Visiting the London School of Tropical Medicine in the wake of the post-World War I influenza pandemic, he was shown the work carried out by W. W. C. Topley's team, involving the study of 'model epidemics' in colonies of mice. Within a year, Flexner, with H. L. Amoss, had begun a study of the epidemiology of mouse typhoid[57].

Before he came to the Rockefeller, during a brief interlude at the University of Pennsylvania, Flexner had had an unexpected visitor, met by chance during a visit to Japan. Hideo Noguchi, from a deprived background in northern Japan, had managed to educate himself by somewhat unorthodox means, and to raise sufficient funds to fulfil his ambition to go to the United States[58]. After a brief period spent at the newly opened Serum Institute in Copenhagen[59], Noguchi joined Flexner at the Rockefeller in 1904. Initially excited by the discovery of *Treponema pallidum* by Schaudinn in 1905, Noguchi did some of his most successful work at the institute on syphilis. His later career was marred by work on the cultivation of viruses which subsequently proved to be erroneous. For the last ten years of his life he struggled tirelessly to prove the false conclusion that yellow fever was caused by a spirochaete, named by him *Leptospira icteroides*. A final desperate attempt to justify his hopeless cause led to his death of yellow fever in West Africa in May, 1928, less than a year after Adrian Stokes had succumbed to the same fate after successfully confirming the identity of the fever as a virus disease[60]. Within a span of less than two years at the end of the 1920s, there were four deaths among the intrepid investigators of yellow fever in West Africa, almost thirty years after Reed and Carroll had been congratulated on having solved the puzzle of its aetiology. Control of the disease by development of a vaccine became possible only after Max Theiler in the 1930s was able to develop a mouse-adapted virus, which in turn led to the development of a vaccine[61].

Under Flexner's directorship, the emphasis was always on infectious disease research and its corollary of animal experiments; inevitably, American antivivisectionist societies, encouraged by their European counterparts and reinforced by members of the Baptist Church, took action against the institute from its inception. Their efforts proved to be in vain, and legislation restricting the experimental use of animals was never introduced[62]. In addition to the laboratory animals kept at the institute, the public was encouraged to bring sick animals to the laboratories for tests and diagnosis. In 1910, one such animal presented at the institute was a Plymouth Rock hen with a malignant growth; the scientist in charge was Peyton Rous[63]. Rous was young and enthusiastic, and very much aware of contemporary trends in infectious disease research and in cancer research. With regard to infectious agents, it was the decade of the filterable viruses; the opportunity to test cell-free filtrates in transmission experiments must never be missed. At the same time, the majority of those working in cancer research were dismissive of any suggestions of viral involvement in oncogenesis, and of theories such as those of Amédée Borrel[64]. The filterable agent inducing a fatal leukaemia in chickens discovered by Ellerman and Bang in 1908[65] was disregarded; leukaemia was not yet considered to be a form of cancer. Nevertheless, Rous decided on a comprehensive study, taking all possibilities into account. He was able successfully to demonstrate the transmission in series of a tumour-producing agent with a cell-free filtrate from the Plymouth Rock hen. It was the first retrovirus to be isolated, and Rous was to extend and amplify his studies of chicken sarcomas for the rest of his long working life. For a variety of reasons, recognition of the importance of his results was delayed for nearly half a century[66].

Rous's early work on sarcomas was carried out in Flexner's department in an institute devoted to medical research in its wider sense. During World War I, when America was still largely unaffected by events in Europe in 1915–16, the Rockefeller Institute began to expand and strengthen its involvement in the sciences which, although outside the immediate orbit of human medicine, yet formed an integral part of the framework for the understanding of medical research. In 1916 the institute opened separate departments of animal diseases and parasitology, genetics, and plant pathology[67]. Fifteen years after he had first declined to leave Harvard for the directorship of the newly established Rockefeller Institute, Theobald Smith returned to Princeton as director of the Institute's Department of Animal Diseases.

For Smith, it was a return to the pure research into economically important livestock diseases he had first carried out at the Bureau of Animal Industry in the 1880s. At the Rockefeller Institute he was able to study in depth a condition which presented a considerable threat to America's dairy industry: contagious abortion in cattle. At a time when *Brucella* infections were subject to study in a number of laboratories in the western world – Bruce had published his first paper on the aetiology of Malta fever in 1887[68] – Smith had become interested through his work on bovine tuberculosis. In 1912 he had published two papers on Bang's *Bacillus abortus*, and had warned of the possibility of milk-borne transmission to man from infected cows[69]. At the Rockefeller, he began to study contagious abortion in bovines on a wider basis; together with work on tuberculosis and the parasites of blackhead in turkeys, this remained a major research interest throughout Smith's time at the department. In the year of his death, the university press published his classic text on parasitism and disease[70].

When at the age of 70 Smith officially 'retired' to become Emeritus Director in 1929, younger men were already engaged in a number of quite different investigations within the Institute. Richard Shope came to the Institute as a recent graduate from medical school in 1925; pending the Institute's translation into a university he became professor of animal pathology in 1952[71]. Two decades after Rous's original work on (retro)virus-induced chicken sarcomas, its results then still doubted by a majority of workers in the field, Shope could offer some supporting evidence when tumour-like growths, this time in a freshly shot rabbit, turned up for examination at the Institute. Shope found the condition to be induced, and to be transmissible in rabbits, by a filterable virus[72]. Its relationship to the virus of myxomatosis, and to other viruses causing growths in populations of wild and of domestic rabbits, was to occupy Shope for many years. Together with the work of his Rockefeller colleagues, especially Peyton Rous and T. M. Rivers, it has had lasting importance for this branch of the growing science of virology.

Very different from the tumour viruses, in many respects at the other end of the spectrum of pathogenic viruses, was another agent of enduring interest to Shope: the virus of swine influenza. First described in the United States at the time of the great influenza pandemic in man following World War I, clinical similarities and coincidental appearance led to suggestions of identity of the disease in man and in swine. In the early 1930s Shope made extensive studies of the disease, its virus, and the bacterial commensal *Haemophilus influenzae suis*, which he found to affect the severity of individual cases. In the light of retrospective serological evidence for close similarities between Shope's strain of swine virus and that responsible for the 1918–19 pandemic in man, he suggested the presence or absence of *Haemophilus influenzae* in man as a possible reason for discrepancies in severity between the first and second waves of the 1918–19 outbreak[73].

In contrast to the European institutes which preceded it, the Rockefeller Institute initially had no qualified veterinarians on its staff; of its forty-odd members, associates, assistants, and fellows, the majority possessed medical degrees, and the rest had advanced degrees in chemistry, experimental biology, and pathology[74]. Nevertheless, from the time of his appointment as director, Simon Flexner had emphasised the need for the institute to be involved in the study of animal disease as well as human disease, and Peyton Rous had been associated with the veterinarian F. S. Jones as early as 1913. When the Department of Animal Pathology was established in 1916, Jones moved to Princeton to work with Smith, the first of a number of outstanding veterinarians in that department, among them H. W. Graybill, who collaborated with Theobald Smith in the extensive experimental work, in the laboratory and in the field, on blackhead in turkeys. The opening of the Department of Animal Pathology certainly gave Smith the opportunity to bring in veterinary colleagues such as Ralph B.

Little, and the year 1916 also saw the appointment of the Institute's first Fellow with a purely veterinary background: Ernest W. Smillie[75]. Where Kilborne had vanished from the annals of research after his considerable contribution to the Texas fever investigations, a new generation of veterinarians went to work in their own right in the Department of Animal Pathology.

The creation of the department was a natural enough development in view of the Institute's, and its Directors', stated aims and intentions. In the circumstances, urgency was added to previously unformulated plans by an outbreak of hog cholera seriously affecting the economy of a number of Western States, and an offer of outside support, later rescinded, for an investigation of the disease[76]. Over the years, the Princeton department of the Institute widened its scope and embraced comparative medicine in its broadest sense, when plant pathology and genetics were added to its concerns. In the 1920s and 1930s, work of fundamental importance was done on plant virus diseases and the nature of filterable viruses[77]. The achievements of, for example, L. O. Kunkel and of W. M. Stanley, and of Avery, MacLeod and McCarty on the rôle of DNA in the pneumococcal transformation phenomenon[78] were clear indications of the inexorable move towards molecular biology.

After the retirement of Simon Flexner in 1935, the extension of subjects studied at the Institute accelerated; at the same time, a number of institutes, modelled on its prototype and in many cases also supported by Rockefeller funds, had been established across the United States and around the world. The Institute had lost its unique status. In such a climate, and following World War II, the Institute's Trustees undertook a major reappraisal of its aims and possibilities in the changed world of the 1950s. Recognising the need to pursue excellence in both research and teaching, the Institute became a graduate university[79].

Since its inception, the Rockefeller Institute had spearheaded pure research in the medical sciences, albeit working closely with a hospital of its own[80], on foundations laid in Europe, and at the Johns Hopkins. Its creation heralded an unprecedented injection of private fortunes, especially through the Rockefeller Foundation, into research both in America and abroad. After establishing the Institute in 1901, and the General Education Board in 1903, the elder Rockefeller and F. T. Gates had the Rockefeller Foundation incorporated in 1913, and the International Health Board in the same year[81]. From then on, the emphasis within the Foundation was on the application of scientific medicine to the prevention of disease, and to public health in a wider sense. The first, and most seminal, manifestation of the new policy was the opening of the Johns Hopkins School of Hygiene and Public Health in 1916; followed, after a far more complex and long-drawn-out period of gestation, by the transformation of the London School of Tropical Medicine into the London School of Hygiene and Tropical Medicine.

The Johns Hopkins School in Baltimore and the London School were the principal centres for the Rockefeller Foundation's International Health Board's campaign for world-wide improvements in public health through applied medical science and education. Between the wars, the Foundation was instrumental in establishing smaller schools and institutes, in both North and South America, throughout Europe, in the Philippines, and on the Indian subcontinent[82].

Among the late additions to this Rockefeller-sponsored crop of institutions dedicated to the pursuit of public health was one situated just below the Arctic Circle, at such northern latitudes (between 63° and 67°) that it could lay no claim to a pressing need for investigation of tropical disease and hygiene. On the other hand, its need to study diseases of domestic animals, in their own right and on a comparative basis, had always been paramount. Furthermore, as a small and isolated island community, Iceland offered ideal conditions for the study of epidemiology, especially of imported diseases of animals and man. Traditionally, diseases of sheep in a nation heavily dependent on sheep farming, had been a common concern of pathologists of all persuasions, whether medical or veterinary. In the 1930s, there were simultaneous outbreaks of three sheep diseases of unknown aetiology on a number of Icelandic farms. Two of the diseases, *visna* ('wasting disease') and *maedi* ('dyspnoea', a respiratory disorder), were later found to be of identical viral origin, although clinically and pathologically distinct. The diseases were apparently introduced with sheep of a special breed, imported by the Icelandic government from Germany in 1933. The animals had been duly placed in quarantine for a period of some weeks on arrival; what nobody had suspected was the introduction of diseases with an incubation period of from one to three years or more. The lack of suspicion is understandable; there was no evidence of prevalence of such diseases in sheep in Germany, where the infections were presumably silent. The destructive effect in Icelandic sheep, geographically protected from contact with sheep populations abroad for centuries, merely illustrates the well-established epidemiological hazards of exposure of virgin populations to previously unencountered pathogens[83].

On the other hand, Icelandic sheep had for years been subject to a slowly developing infection with an unusually long incubation period. Like *visna*, *rida* is a disease of the central nervous system; it is in fact the Icelandic form of scrapie. The simultaneous presence of these three slowly developing diseases and a fourth, *jaagsiekte*, gave rise to an extensive study with lasting results, both scientific and practical, in the pathology department of Reykjavik's medical school[84].

In 1941, when science and scientists on the European mainland were effectively cut off from visits to North America by the war, Iceland, by then a sovereign state sharing only the person of the monarch with Denmark, was unaffected by the German oc-cupation. Instead the island received protection from British, and after 1941 from American forces; closer links developed with the United States. This facilitated scientific exchanges, and between 1941 and 1943 Björn Sigurdsson, a medical graduate and associate of Niels Dungal, the Reykjavik professor of pathology, worked as a visiting Fellow at the Rockefeller Institute on the development of methods of cultivation of viruses in chick embryos[85]. By the time he left the Institute, he had impressed the resident staff, and the Rockefeller Foundation was to consider favourably an application made in 1944 by the Icelandic Government, which in 1941 had acquired an old farm, Keldur, hoping to turn it into an institute for experimental pathology. Over the next few years, the Rockefeller Foundation contributed $200 000, coincidentally the same amount first granted for merely planning the Rockefeller Institute at the beginning of the century. In Reykjavik, the Icelandic Government met the building costs, and the $200 000 covered the escalating costs of modern equipment. Towards the end of 1948, the first buildings at Keldur had been completed, and work could begin in the Institute. It was the northernmost of all the institutes inspired by the Rockefeller Institute, and supported in part by Rockefeller funds; and it was the only one whose main preoccupation was diseases of sheep. By 1949, Sigurdsson's work was published from the Institute for Experimental Pathology at Keldur, where it still flourishes as a major establishment of international repute, albeit plagued by the financial difficulties so prominent in the academic world of the late twentieth century. Reflecting its traditional theme of sheep diseases, the institute's staff is composed of medical and veterinary graduates in about equal proportions. Without a veterinary school of its own, Iceland has had over the years to rely on training abroad for its veterinarians, mostly in other Scandinavian countries[86].

Before his early death at the age of 46, it was his work at Keldur which led Sigurdsson to formulate the concept of 'slow virus diseases', which has since been growing in importance and complexity. He first explained his ideas publicly in a series of Special University Lectures given at the University of London in March, 1954[87]. Central to his arguments were the behaviour and epidemiology of the diseases '*rida*' and '*maedi*' as observed in Iceland; and he made a point of distinguishing between 'slow' infections, which have a protracted incubation period before developing an unvarying pattern with a fatal outcome, and the 'chronic' ones, e.g. tuberculosis, syphilis, malaria, all exhibiting an irregular course, and an unpredictable outcome: the patient may recover, or may die of the disease after a shorter or a very long period of living with the disease[88].

Initially, Sigurdsson tentatively included in the group of 'slow virus infections' those malignant neoplastic animal diseases which had only shortly before been shown to be caused by filterable agents. That interpretation has long since been overtaken by the results of extensive work on the molecular biology of oncogenic viruses. On the other hand, the realisation that the agents of scrapie, and of a number of related diseases of animals and man, although filterable, have other properties not reconcilable with the accepted definition of the conventional filterable viruses, has caused taxonomic confusion and growing anxiety over public health issues. The recent furore caused in Britain and among her

European partners by the appearance of bovine spongiform encephalopathy (BSE) – 'mad cow disease' – may have been accentuated by the power of the media in the late twentieth century. Yet there is a chilling analogy with the concern over the safety of meat and milk at the height of the mid-nineteenth century outbreak in Britain of rinderpest, when knowledge of the pathogenesis of that affliction was as scant as is that of scrapie and the related diseases today.

For although increased knowledge and understanding have removed the animal cancers originally included by Sigurdsson from today's list of scrapie-related diseases, others have been added in the last thirty years. Until the recent inclusion of BSE, thought to be scrapie transmitted to cattle with contaminated feedstuffs, only one other scrapie-related disease in animals had been observed: a transmissible encephalopathy of farmed mink in North America. First observed in 1947, it remains a rare disease which does not maintain itself in a given population of mink. The fact that there is strong circumstantial evidence for initial introduction of the agent by feeding mink with scrapie-contaminated tissue, may cause surprise that decades later no attempts were made to prevent the use of similarly contaminated tissue in cattle feed in Europe[89].

In addition to scrapie, *rida*, BSE and transmissible mink encephalopathy, two diseases of man, characterised by slow progressive degeneration of the central nervous system, appear to be caused by similar unconventional agents. One, kuru, was discovered in 1955 and was confined to a limited population of intermarrying linguistic groups in the highlands of Papua New Guinea. The other, Creutzfeldt–Jakob disease (CJD), was first described in 1920 as a degenerative disease of man. Kuru was easily contained once the Fore people were persuaded to discontinue the practice of ritual endocannibalism as a rite of mourning and a mark of respect for dead relatives[90]. In contrast, CJD occurs world-wide, in both sexes and among different races. Epidemiological surveys have failed to establish any link between scrapie infection in sheep, and incidence of CJD in those professionally occupied with such sheep. On the other hand, some evidence has been found of familial patterns of inherited susceptibility to the disease[91].

The above epidemiological findings offer scant support for current public fears and anxieties over possible transmission to man in the wake of the appearance of BSE in cattle fed on scrapie-contaminated products. Nevertheless, so long as the very nature of the agents of these diseases remains unexplained and a subject of controversy between scientists of opposing views with different theories, it would be foolish to exert anything but constant vigilance over public health aspects of the problems of diseases in animals and man caused by unconventional – or conventional – agents.

Over the past three centuries, studies of infectious diseases of animals have been pursued in the interests of the animals themselves as well as of man. From the tentative conclusions concerning the nature and epidemiology of the cattle plagues of the eighteenth century reached by Ramazzini and by Lancisi, to the as yet unresolved problems of the nature and epidemiology of scrapie, BSE, and Creutzfeldt–Jakob disease now under scrutiny at the molecular level, the lessons to be learnt by veterinarians and comparative pathologists have lost none of their importance and immediacy. As experiments in tissue culture systems increasingly replace more controversial animal experiments, comparative medicine continues to prove its worth in ever more sophisticated ways.

Notes

CHAPTER I

1. Raoul Caveribert, *La vie et l'oeuvre de Rayer (1793–1867)*, Paris, M. Vigne, 1931, pp. 15–19. *Cf.* chapter 7.

2. *The Cambridge Ancient History*, 3rd ed., vol. I part 1, London, Cambridge University Press, 1970; Felix Regnault, 'La paleopathologie et la médecine dans la préhistoire', in: *Histoire générale de la médecine*, Laignel-Lavastine (ed.), Paris, A. Michel, 1936, vol. I; E. Leclainche, *Histoire de la médecine vétérinaire*, Toulouse, Office du Livre, 1936.

3. Calvin W. Schwabe, *Cattle, priests and progress in medicine*, Minneapolis, University of Minnesota Press, 1978, pp. 9–10.

4. The possible interactions between early man, early animal herds, and parasites and micro-organisms have been discussed by W. H. McNeill in *Plagues and peoples*, Oxford, Basil Blackwell, 1976, chapter 2. On the question of the impact of animal disease as such, McNeill is silent.

5. M. Stöber, 'Überblick über die Geschichte des Buiatrik', *Berl. Münch. tierärztl. Wschr.*, 1965, 78: 461–5.

6. George Fleming, *Animal Plagues*, London, Chapman and Hall, 1871, p. xxv.

7. C.-E. A. Winslow, *The conquest of epidemic disease*, Princeton, New Jersey, Princeton University Press, 1943, chapters I and II.

8. Sir Frederick Smith, *The early history of veterinary literature*, London, J. A. Allen and Co., 1976, pp. 5–6a.

9. F. L. Griffith, *The Petrie Papyri*, London, Quaritch, 1898, pp. 12–14; C. W. Schwabe, 'An integrated approach to the biomedical implications of ritual bull sacrifice in ancient Egypt', Paper given at Wellcome Symposium, London, June, 1989.

10. O. P. Jaggi, *Scientists of ancient India and their achievements*, Delhi, Atma Ram, 1966, pp. 121–5; *idem, Indian system of medicine*, Delhi, Atma Ram, 1973, pp. 199–203.

11. G. Soulie de Moraut, 'Chine et Japon' *in*: Laignel Lavastine, *op. cit.* note 2, pp. 527–54; J. Smithcors, *The evolution of the veterinary art*, Kansas City, Veterinary Medicine Publishing Co., 1957.

12. Leclainche, *op. cit.*, note 2.

13. G. M. Lancisi, *Dissertationis historicae de bovilla peste*, Rome, J. N. Salvioni, 1715, p. 182.
14. *Hippocratic writings*, Penguin Classics, 1983, pp. 87–138 and 148–169.
15. The details of these confusions are discussed by Smith, *op. cit.* note 8, pp. 33–4 and 71–4.
16. D'Arcy W. Thompson, *The works of Aristotle translated into English*, vol. IV, 'Historia Animalium', book VIII. The translator's notes warn that book VIII shows signs of 'an alien hand' and 'interpolation' by later 'negligent scribes'. Frequent transcriptions, translations and re-translations of such early works have provided ample opportunity for obscuring the language of the original manuscripts.
17. In animals, three forms of the disease are recognised: an acute, a sub-acute, and a per-acute or apoplectic type. Difference in manifestations of anthrax between different species of higher animals has sometimes led to confusion with rabies; George Fleming in *Rabies and Hydrophobia*, London, Chapman and Hall, 1872, felt the need to include a chapter on 'Analogies and dissimilarities between rabies and anthrax'.
18. See Hieronymi Fracastorii, *De contagione et contagiosis morbis et eorum curatione*, translation and notes by Wilmer Cave Wright, New York and London, P. G. Putnam's Sons, 1930, pp. 128–9. Fracastoro believed that Aristotle was drawing a distinction between animals, which inevitably developed the disease when bitten by a rabid dog, and man, who might or might not develop clinical symptoms.
19. J. F. Dobson, 'Herophilus of Alexandria', *Proc. R. Soc. Med.*, 1925, 18: 19–32; and, *idem*, 'Erasistratus', *ibid.*, 1927, 20: 825–32.
20. Celsus, *De medicina*, with an English translation by W. G. Spencer, London, W. Heinemann, 1938; on rabies, book V, pp. 112–15; on Herophilus and Erasistratus, prooemium, p. 15, and books III–VII, *passim*.
21. Charles Singer, 'Galen as a modern', *Proc. R. Soc. Med.*, 1949, 42: 563–70; W. L. H. Duckworth, 'Some notes on Galen's anatomy', Linacre Lecture 6 May 1948, Cambridge, Heffer and Sons Ltd., 1949; Vivian Nutton (ed.), *Galen: problems and prospects*, London, Wellcome Inst. Hist. Med., 1981.
22. F. Smith, *op. cit.* note 8, p. 12–13.
23. *ibid.*, and Smithcors, *op. cit.* note 11, esp. pp. 80–1.
24. William Bulloch, *The history of bacteriology*, Oxford University Press, 1938, chapter 1.
25. Saul Jarcho, 'Medical and non medical comments on Cato and Varro, with historical observations on the concept of infection', *Trans. Stud. Coll. Phys. Phila.*, ser. 4, 1976, 43: 372–8.
26. *Lexicon der Alten Welt*, Zürich and Stuttgart, Artemis Verlag, 1965.
27. Valentino Chiodi, *Storia della veterinaria*, Milan, Farmitalia, 1957, pp. 128–33.
28. F. Smith, *op. cit.* note 8, pp. 30–6.
29. Fleming, *Animal Plagues*, *op. cit.* note 6, p. 31.

30. Vegetii Renati, *Artis veterinariae, sive mulomedicinae*, Basel, J. Faber, 1528, p. 39.
31. J. M. Roberts, *The Hutchinson History of the World*, London, Hutchinson and Co. Ltd., 1976, pp. 453–60.

CHAPTER 2

1. Rhazes, 'A treatise on the small-pox and measles', *Med. Classics*, 1939, 4 (i): 22–84; see pp. 24–7.
2. O. Cameron Gruner, *A treatise on the Canon of Medicine in Avicenna*, London, Luzac & Co., 1930, p. 171.
3. Albertus Magnus, *De animalibus*, Venice, Joannes & Gregorius de Gregorio, 1495, 217a.
4. *op. cit.* note 2.
5. D. Abulafia, *Frederick II: A medieval emperor*, London, Allen Lane, Penguin Press, 1988; R. Froehner, *Kulturgeschichte der Tierheilkunde*, Konstanz, Terra-Verlag, 1968, vol. 3.
6. V. Chiodi, *Storia della veterinaria*, Milano, Farmitalia, 1957.
7. Lorenzo Rusio, *Opera de l'arte del mascalcio*, Venice, M. Tramezino, 1543; *La mascalcia di Lorenzo Rusio*, ed. Pietro Delprato, Bologna, Presso G. Romagnoli, 1867, vol. I, pp. 355–9.
8. C. E. A. Winslow, *The conquest of epidemic disease: a chapter in the history of ideas*, Princeton University Press, 1943; see chapter 6.
9. *La Grande Chirurgie* de Guy de Chauliac, ed. E. Nicaise, Paris, Germier Baillière et Cie, 1890.
10. Vivian Nutton, 'The seeds of disease: an explanation of contagion and infection from the Greeks to the Renaissance', *Med. Hist.*, 1983, 27: 1–34.
11. Hieronymi Fracastorii, *op. cit.*, chapter I, note 14.
12. Winslow, *op. cit.* note 8.
13. Athanasius Kircher, *Scrutinium physico-medicum contagiosae luis, quae pestis dicitur*, Rome, Masardus, 1658.
14. G. Fleming, *Animal Plagues*, *op. cit.* chapter 1, note 5.
15. Agostino Gallo, *Le vinti giornate dell'agricoltura*, Turin, Beuilacqua, 1580.
16. Charles Webster, *The great instauration: science, medicine and reform 1626–1660*, London, Duckworth, 1975.
17. Richard Lower, 'The Method observed in Transfusing the bloud out of one animal into another', *Phil. Trans. R. Soc. Lond.*, 1666, 1: 353–8; *idem*, 'An account of the Experiment of Transfusion, practised upon a Man in London', *ibid.*, 1667, 2: 557–9; 'An Extract of a Letter, Written from Dantzick to the Hon. R. Boyle, containing the success of some Experiments of Infusing Medicines into humane Veines', *ibid.*, 1668, 3: 766–7.
18. 'An account of Mr. Richard Lower's newly published Vindication of Doctor Willis's Diatriba de Febribus', *ibid.*, 1665, 1: 77–8; 'An account of Dr. Sydenham's Book, entitled, *Methodus Curandi Febres, Propriis*

observationibus superstructa', *ibid*., 1666, 1: 210–13; Account of Nathaniel Hodges' history of the London plague of 1665, *ibid*., 1672, 7: 4028–30; Account of Thomas Willis' *De Anima Brutorum*, *ibid*.: 4071–3.

19. *The Record of the Royal Society of London*, 4th ed., London, Royal Society, 1940.

20. F. Slare, 'An account of a Murren in Switzerland, and the Method of its cure', *Phil. Trans. R. Soc. Lond.*, 1683, 13: 93–5.

21. Jacques Labessie de Solleysel, *Le parfait maréschal*, Paris, G. Clousier, 1679. The quote is rendered in the colourful translation by Sir William Hope, *The compleat horseman*, London, Gilliflower, 1696.

CHAPTER 3

1. *The works of Thomas Sydenham, M.D.*, translated from the Latin edition of Dr. Greenhill with a life of the author by R. C. Latham, London, Sydenham Society, 1848, vol. 1, pp. 123–51; 187–92; 218–25.

2. Donald R. Hopkins, *Princes and peasants. Smallpox in history*, Chicago and London, University of Chicago Press, 1983, pp. 37–41.

3. Charles Creighton, *A history of epidemics in Britain*, Cambridge University Press, 1891–4, vol. 2, pp. 471–5.

4. Emanuel Timonius, 'An account, or history, of the procuring the smallpox by incision, or inoculation, as it has for some time been practised at Constantinople', *Phil. Trans. R. Soc. Lond.*, 1714, 29: 72–82.

5. Jacobus Pylarinus, *Nova et tuta Variolas excitandi per Trans-plantationem methodus*, Venice, Gabriel Hertz, 1715.

6. Genevieve Miller, *The adoption of inoculation for smallpox in England and France*, Philadelphia, University of Pennsylvania Press, 1957.

7. See Ludvig Hektoen, 'Experimental measles', *J. infect. Dis.*, 1905, 2: 238–55; and on rinderpest, *cf.* the work of D. P. Layard described in the following chapter, pp. 58–9.

8. F. Toggia, *Storia e cura delle più familiari malattie dei buoi analoghe a quelle de cavallo*, quoted by Fleming, vol. 1, p. 170, and Chiodi, p. 332.

9. See, e.g., C. F. Mullett, 'The cattle distemper in mid-eighteenth century England', *Agric. Hist.*, 1946, 20: 144–65; and also C. F. Cogrossi, note 17 below.

10. See John Gamgee, *The cattle plague*, London, Robert Hardwicke, 1866, pp. 288–9.

11. For the state of epidemiology two centuries later, in the nineteenth century before acceptance of the germ theory, see W. Coleman, *Yellow fever in the north*, Madison, University of Wisconsin Press, 1987.

12. Bernardino Ramazzini, *Constitutionem epidemicarum Mutinensium*, Padua, J. B. Conzatti, 1714, esp. pp. 19–23 and 135–40; quoted by Fleming, *op. cit.*, pp. 158–65, passim.

13. *idem, De contagiosa epidemia, quae in Patavino agro, & tota fere Veneta ditione in boves irrepsit*, Padua, Conzatti, 1712; see also Wilmer

Cave Wright's introduction to her translation of Ramazzini's *Diseases of workers*, New York, Hafner, 1964; and John M. McDonald, 'Ramazzini's dissertation on rinderpest', *Bull. Hist. Med.*, 1942, 12: 529–39.

14. Giovanni Maria Lancisi, *De motu cordis et aneurysmatibus*. Opus posthumum, Rome, M. Salvioni, 1728.

15. *idem, Dissertatione historicae de bovilla peste*, Rome, J. M. Salvioni, 1715; see also L. Wilkinson, 'Rinderpest and mainstream infectious disease concepts in the eighteenth century', *Med. Hist.*, 1984, 28: 129–150.

16. *idem, De noxiis paludum effluviis eorumque remediis*, Rome, Salvioni, 1717.

17. C. F. Cogrossi, *Nuova idea del male contagioso de' buoi'*, Milan, 1714; facsimile edition with English translation by D. M. Schullian and a foreword by Luigi Belloni, for Sezione Lombarda della società italiana di microbiologia, 1953.

18. *Osservazioni intorno a' pellicelli del corpo umano fatte dal Dottor Gio: Cosimo Bonomo, e da lui con altre osservazioni scritte in una lettera all' Illustriss. Sig. Francesco Redi*, Firenze, 1687. The observations were made jointly by the pharmacist Giacinto Cestoni (1637–1718) and Bonomo (*d.* 1696).

19. See, e.g., E. Duclaux, *Traité de microbiologie*, vol. 1, Paris, 1898, pp. 34–35; Adam Neale, *Researches to establish the truth of the Linnean doctrine of animate contagions*, London, 1831 (*cf.* also chapter 8), pp. 7–10; R. Friedman, *The story of scabies*, New York, Froben Press, 1948, vol. 1.

20. G. M. Lancisi, 'Epistolaris dissertatio ad Doctissimum virum Antonium Mariam Borromaeum de bovilla peste', in the historical dissertation, note 15 above, pp. 179–81.

21. Frank N. Egerton III, 'Richard Bradley's illicit excursion into medical practice in 1714', *Med. Hist.*, 1970, 14: 53–62.

22. Among Bradley's works see especially *The Plague at Marseilles consider'd*, London, W. Mears, so popular it went through four editions all in 1721; *Precautions against infection*, London, Thomas Corbett, [1722]; *New improvements of planting and gardening, both philosophical and practical*, London, W. Mears, 1717–18; the 7th edition of the latter was published in 1739.

23. Humbert Mollière, *Un precurseur Lyonnais des théories microbiennes. J.-B. Goiffon et la nature de la peste*, Bale-Lyon-Genève, Henri Georg, [1886]; M. Lannois, 'Jean-Baptiste Goiffon 1658–1730', *Hippocrate*, 1937, 5: 193–206.

24. Charles Singer, 'Benjamin Marten, a neglected predecessor of Louis Pasteur', *Janus*, 1991, 16: 81–98; Benjamin Marten, *A new theory of consumptions*, London, R. Knaplock, 1720; Raymond Williamson, 'The germ theory of disease. Neglected precursors of Louis Pasteur (Richard Bradley, Benjamin Marten, Jean-Baptiste Goiffon)', *Ann. Sci.*, 1955, 11: 44–57.

25. N. Andry, *De la génération des vers dans le corps de l'homme*, Paris, L. d'Houry, 1700.

26. On the 'innate seed' concept in smallpox see G. Miller, note 6 above.

27. Thomas Fuller, *Exanthematologia: or, an attempt to give a rational account of eruptive fevers, especially of measles and smallpox*, London, C. Rivington and S. Austen, 1730.

28. Cecil Wall, *The history of the surgeons' company 1745–1800*, London, Hutchinson's, [1937].

CHAPTER 4

1. John Gamgee, *The Cattle Plague*, London, Hardwicke, 1866, pp. 285–335; Hans Chr. Bendixen, *Den kongelige Veterinaerskole*, Copenhagen, 1973, chapter 3.

2. The year 1714 saw not only a change of monarch, but also a resumption of Whig – and Robert Walpole's – dominance after a brief spell of Tory supremacy.

3. 'An extract from the *Acta Eruditorum* for the month of March, 1713. Pag. III', *Phil. Trans. R. Soc. Lond.*, 1714, **29**: 46–9.

4. William Gibson, *A new treatise on the diseases of horses*, London, A. Millar, 1751.

5. Thomas Bates, *An Enchiridion of Fevers incident to Sea-Men in the Mediterranean*, London, John Barns, 1709.

6. Thomas Bates, 'A brief account of the contagious disease which raged among the milch cowes near London, in the year 1714. And of the methods that were taken for suppressing it', *Phil. Trans. R. Soc. Lond.*, 1718, **30**: 872–85. Bates was elected FRS in 1718. The political decisions and reports of the Justices of the Peace are in the Calendar of Treasury Papers 1714–19, Vol. CLXXXII, pp. 30–5.

7. e.g. 'A Recipe: or the ingredients of a medicine for the spreading mortal distemper amongst cows: lately sent over from Holland, where a like distemper raged amongst the Black Cattle', *Phil. Trans. R. Soc. Lond.*, 1714, **29**: 50.

8. Letter in *Gent.'s Mag.*, 1744, **14**: 585–8 [Nov. 1744]; Thomas Bates, *ibid.*, 1745, **15**: 528.

9. Cromwell Mortimer, 'A Third Account of the Distemper among the Cows', *Phil. Trans. R. Soc. Lond.*, 1746, **44**: 4–10; see especially Appendix, p. 7; also reported *verbatim* in *Gent.'s Mag.*, 1747, **17**: 55. The two white calves are also mentioned by Layard, see note 20 below, p. 12.

10. *Gent.'s Mag.*, 1745, **15**: 630–1.

11. Daniel Peter Layard, 'A Discourse on the Usefulness of Inoculation of the horned Cattle to prevent the contagious Distemper among them', *Phil. Trans. R. Soc. Lond.*, 1758, **50**: 528–38, especially pp. 530–1.

12. See Charles F. Mullett, 'The cattle distemper in mid-eighteenth century England', *Agric. Hist.*, 1946, **20**: 144–65; also *Gentleman's Magazine*, and *London Magazine*, for the years 1744–59, *passim*.

13. Cromwell Mortimer, 'Some Account of the Distemper raging among the Cow-kind in the neighbourhood of London, together with some remedies proposed for their recovery', *Phil. Trans. R. Soc. Lond.*, 1745, 43: 532–7.

14. Theophilus Lobb, *A treatise on the small pox.* The 2nd edition, corrected, with large additions, London, T. Woodward and C. Davis, 1741.

15. Theophilus Lobb, *Letters relating to the plague, and other contagious distempers*, London, J. Buckland, 1745.

16. F. J. Billeskov Jansen, 'Kvaegsyge og Vikingetogter Ludvig Holbergs to Bidrag til Videnskabernes Selskabs Skrifter', in: *Ludvig Holbergs to Bidrag til Videnskabernes Selskabs Skrifter*, Copenhagen, Kgl. Danske Videnskabernes Selskab, 1972; Hans Rieck, 'Studien zu Betrachtungen der Kopenhagener Professoren J. B. von Buchwald, Georg Detharding und Ludwig von Holberg zur Rinderpest 1745 in Dänemark', *Hist. med. vet.*, 1980, 5 (3): 69–70.

17. Ludvig Holberg, *b.* Norway 1684, *d.* 1754; prolific writer on history and philosophy who introduced the commedia dell'arte form to Danish eighteenth century theatre.

18. Ludvig Holberg, 'Korte Betaenkning over den nu regierende Kvaeg-Syge med nogle oeconomiske Anmerkninger', *Videnskabernes Selskabs Skrifter*, 1745, 2: 385–402, especially pp. 386–8, 390 and 400–1. Facsimile in above, note 16.

19. Thomas Bates, note 8.

20. D. P. Layard, *An essay on the nature, causes and cure of the contagious distemper among the horned cattle in these kingdoms*, London, Rivington, 1757, see introduction, p. 8.

21. *ibid.*, chapter VII, 'On inoculation', pp. 100–10.

22. D. P. Layard, 'A Letter to Joseph Banks,... relative to the Distemper among the horned Cattle', *Phil. Trans. R. Soc. Lond.*, 1780, 70: 536–45.

23. Johs. Kristiansen, 'Almindelig Anordning imod Hornqvaegets Sygdom, for Kongeriget Danmark', *Dansk veterinaerhistorisk Årbog*, 1988, 33: 76–86; including facsimile of the decree, dated Christiansborg Castle, 30 Nov., 1778.

24. See, e.g. A. van der Schaaf, 'Geert Reinders (1737–1835)', *Hist. vet. med.*, 1978, 3: 89–98.

25. *Oeuvres de Pierre Camper qui ont pour objet l'histoire naturelle, la physiologie et l'anatomie comparée*, 3 vols, Paris, H. J. Jansen, 1803. Vol. 3 contains 'Lecons sur l'épizootie qui règna dans la province de Groningen en 1769', pp. 17–213. In vol. 2 of the *Oeuvres* is another example of Camper's abiding interest in comparative approaches: 'Réponse à la question proposée en 1783 par la Société Batave de Rotterdam: "Exposer les raisons physiques pourquoi l'homme est sujet à plus de maladies que les autres animaux? Quels sont les moyens de rétablir sa santé, qu'on peut emprunter des observations que fournit l'anatomie comparée?"' (pp. 283–448).

26. See V. Nutton, 'The seeds of disease: an explanation of contagion and

infection from the Greeks to the Renaissance', *Med. Hist.*, 1983, 27: 1–34.

27. M. A. Plenciz, *Opera medico physica*, Vienna, J. I. Trattner, 1762, especially pp. 30–44; 140–5; and in the *Additamentum* on 'Lues bovina', p. 199–217.

28. George Fleming, *Animal Plagues*, London, Chapman Hall, 1871.

29. Albrecht Haller, *Mémoire sur la contagion parmi le betail. Abhandlung von der Viehseuche*, Bern, 1773. An English translation of most of the relevant passages from this treatise may be found in Fleming's *Animal Plagues* above, pp. 446–60, together with Fleming's comments.

30. D. Barberet, *Mémoire sur les maladies épidémiques des bestiaux*, Paris, Veuve d'Houry, 1766; also included in: *Receuil de mémoires et observations-pratiques sur l'épizootie*, Lyon, Reymann, 1808.

31. Felix Vicq d'Azyr, *Exposé des moyens curatifs et préservatifs que peuvent être employés contre les maladies pestilentielles des bêtes à cornes*, Paris, Mérigot, 1776.

32. Dossie's 'Observations on the murrain or pestilential disease of meat cattle', in: *Memoirs of Agriculture*, vol. 2, 1771, are reproduced by George Fleming, note 28 above, pp. 307–57. See also F. W. Gibbs, 'Robert Dossie (1717–1777) and the Society of Arts', *Ann. Sci.*, 1951, 7: 149–72.

33. C. Linné, *Systema naturae*, Vienna, Trattner, 1767, vol. I, part 2, see pp. 1326–7. The class 'Chaos', including 'Chaos infusorium', was placed under 'Vermes'. This classification went through numerous editions published in Linnaeus' lifetime, but had been removed when an English translation, *A general system of nature*, was published in London in 1806.

34. Erasmus Darwin, *Zoonomia: or, the laws of organic life,* vol. II, London, J. Johnson, 1796, p. 249.

CHAPTER 5

1. E. Leclainche, *Histoire de la médecine vétérinaire*, Toulouse, Office du Livre, 1936; Henri Hours, *La lutte contre les épizooties et l'école vétérinaire de Lyon au XVIIIᵉ siècle*, Paris, Presses Universitaires de France, 1957; A. Railliet et L. Moulé, *Histoire de l'Ecole d'Alfort*, Paris, Asselin et Houzeau, 1908.

2. Marc Mammerickx, *Claude Bourgelat Avocat des vétérinaires*, Brussels, for the author, 1971.

3. H. Hours, note 1 above, p. 24.

4. C. Bourgelat, *Nouveau Newkastle ou Traité de cavalerie*, 1740, which Leclainche, note 1 above, calls 'conscientious but devoid of originality'; *idem, Eléments d'hippiatrique, ou nouveaux principes sur la connoissance et sur la médecine des chevaux...*, Lyons, H. Declaustie et Frer, Duplain, 1750–3.

5. 'Les difficultés de l'école de Lyon', in: H. Hours, note 1 above, pp. 28–44; also Table I, p. 27.

6. 'L'art vétérinaire à la recherche d'une position logique', *ibid.*, pp. 61–72.

7. H. C. Bendixen, 'The Royal Veterinary School in Copenhagen. Highlights from the time of Peter Chr. Abildgaard and Erik Nissen Viborg', *Hist. Med. Vet.*, 1976, 1: 70–7; *idem, Omkring 200-året for oprettelsen af Den Kongelige Veterinaerskole*, Copenhagen, 1973, p. 67: transcript of an Abildgaard MS in the Danish State Archives.

8. Correspondence reproduced in Leclainche, pp. 243–4.

9. Etienne Lafosse in the preface to his *Cours d'Hippiatrique*; quoted *verbatim* by Leclainche, p. 242.

10. The book by the elder Lafosse is: E. G. Lafosse, *Traité sur le véritable siège de la morve des chevaux, et les moyens d'y remédier*, Paris, David et Gonichon, 1749. Details of the acrimonious relationship between the Lafosses and Bourgelat may be found in: C. H. Eby, 'Lafosse and his book', *Record (Friends of the Library of Washington State University)*, 1960, pp. 39–43, and also in Leclainche, note 1. Further details of the study of glanders in France and elsewhere are in L. Wilkinson, 'Glanders: medicine and veterinary medicine in common pursuit of a contagious disease', *Med. Hist.*, 1981, 25: 363–84.

11. 'Bertin (Henri-Léonard-Jean-Baptiste)' in: *Nouvelle Biographie Générale*, Paris, Firmin Didot-Frères, 1866; see also Caroline C. Hannaway, 'Veterinary medicine and rural health care in pre-revolutionary France', *Bull. Hist. Med.*, 1977, 51: 431–47.

12. J. Bost, 'Les écoles vétérinaires francaises (Lyon et Alfort) face aux épizooties du XVIIIème siècle' in: *Histoire des grandes maladies infectieuses*, Cycle des Conférences Année 1979–1980, Institut d'Histoire de la Médecine, Université Claude Bernard, Lyon, see pp. 143–50.

13. H. Hours, *op. cit.* note 6, p. 69.

14. Donald R. Hopkins, *Princes and peasants. Smallpox in history*, Chicago and London, University of Chicago Press, 1983, pp. 70–2.

15. Caroline C. Hannaway, 'The Société Royale de Médecine and epidemics in the Ancien Régime', *Bull. Hist. Med.*, 1972, 46: 257–73; J. Noir, 'Un savant, un innovateur, un réalisateur. Felix Vicq d'Azyr (1748–1794)', Extraits du *Concours Médical* des 6 avril et 4 mai, Clermont, Thiron, 1927: 1–23; Léon Moulé, 'Vicq d'Azyr et la pathologie animale', *Bull. Soc. Hist. Méd.*, 1923, 17: 192–205.

16. H. Hours, *op. cit.* note 6, see p. 66.

17. Vicq d'Azyr et Jeanroi, 'Rapport fait à la Société Royale de Médecine au sujet de l'épidémie qui a regné à Villeneuve-les-Avignon', *Histoire de la Société Royale de Médecine*, 1776, 1: 213–25.

18. Arthur Young, *Travels in France during the years 1787, 1788 and 1789*, ed. Constantia Maxwell, Cambridge University Press, 1929, p. 87.

19. J. Noir, *op. cit.* note 15, see p. 11.

20. *cf.* Caroline C. Hannaway, *op. cit.* note 11.

21. P. Thillaud, 'Vicq d'Azyr (1748–1794): Anatomie d'une élection', *Hist. sci. méd.*, 1986, 20: 229–36.

22. Sir F. Smith, *The early history of veterinary literature and its British development*, 4 vols, London, J. A. Allen and Co., 1976; on Sainbel see vol. 2, pp. 184–203; Bracy Clark, 'Vial de St. Bel and the early history of the London Veterinary College', *Edinb. vet. rev.*, 1861, 3: 129–37; H. Hours, *op. cit.* note 2, see footnote 6, p. 29.

23. L. P. Pugh, *From farriery to veterinary medicine, 1785–1795*, Cambridge, W. Heffer and Sons, Ltd., 1962; Garry Alder, *Beyond Bokhara. The life of William Moorcroft Asian explorer and pioneer veterinary surgeon 1767–1825*, London, Century Publishing, 1985; Sir F. Smith, *op. cit.* note 22, vol. 3, pp. 3–13.

24. Edward Jenner, *An inquiry into the causes and effects of the variolae vaccinae*, London, Sampson Low, 1798; World Health Organisation, *The global eradication of smallpox*, Geneva, World Health Organisation, 1980.

25. For the history of Jenner's vaccine, and the unsolved riddle of its virus, see Derrick Baxby, *Jenner's smallpox vaccine*, London, Heinemann Educational Books, 1981.

26. E. Jenner, 'Observations on the distemper in dogs', *Med. Chir. Trans.*, 1815 (3rd ed.; 1st ed. 1809), 1: 265–70; G. Fleming, *Animal Plagues*, vol. 1, London, Chapman and Hall, 1871, pp. 410–11.

27. L. Wilkinson, '"The other" John Hunter, M.D., F.R.S. (1754–1809): his contributions to the medical literature, and to the introduction of animal experiments into infectious disease research', *Notes and Records R. Soc. Lond.*, 1982, 36: 227–41.

28. Jessie Dobson, *John Hunter*, Edinb. and London, E. and S. Livingstone Ltd., 1969, pp. 280–1; E. Allen and J. E. Cooper, 'A Hunterian account of human rabies', *Vet. Hist.*, winter 1980/81, n.s. 1: 146–8.

29. J. Hunter, 'Observations, and heads of inquiry, on canine madness, drawn from the cases and materials collected by the Society, respecting that disease', *Trans. Soc. Improv. Med. Chir. Knowl.*, 1793, 1: 194–329.

30. The baron was given to tireless commercial exploitation of his remedy, a not unusual activity among inventors of remedies, spurious or not, in the eighteenth century. For details of his career, see G. W. Schrader, *Thierärztliches Biographisch-literarisches Lexicon*, Stuttgart, Ebner and Seubert, 1863, pp. 399–402. For human equivalents, see Roy Porter, *Health for sale. Quackery in England 1660–1850*, Manchester University Press, 1989.

31. C. Bourgelat, *Matière médicale raisonnée ou précis des médicamens considérés dans leurs effets, à l'usage des élèves de l'Ecole Royale Vétérinaire avec les formules médicinales de la même école*, Lyon, Jean-Marie Bruyset, 1765, see pp. 135–41.

32. E. Viborg, 'Kort Efterretning om Snive, Hestekopper og Quaerke, oplyst ved nyere anstillede Forsøg med disse Sygdoms-Smitter', *Physicalsk-oecon. mediochir. Bibliothek*, December 1795, pp. 330–6.

33. *ibid.*, February 1795, pp. 119–39.
34. S. A. Bardsley, 'Miscellaneous observations on canine and spontaneous hydrophobia: to which is prefixed, the history of a case of hydrophobia, occurring twelve years after the bite of a supposed mad dog', *Mem. Lit. Phil. Soc. Manch.*, 1793, 4: 431–88, esp. p. 483.
35. *idem.*, *Medical reports of cases and experiments, with observations, chiefly derived from hospital practice: to which are added an enquiry into the origin of canine madness; and...a plan for its extirpation from the British Isles*, London, R. Bickerstaff, 1807, esp. pp. 330–5.
36. C. J. Bredin, *Notice biographique sur le professeur Buniva, de Turin*, Paris, Mme Huzard, 1835; H. C. Bendixen, *Omkring 200-året...*, *op. cit.* note 7; Brugnone had studied medicine before attending the veterinary schools.
37. M. F. Buniva, 'Mémoire contenant les plus remarquables notices historiques, et les résultats les plus intéressans de ses observations et experiences, relatives à l'épizootie bos-hongroise qui fait des ravages en Piémont depuis la fin de l'an 1793; *idem*, 'Histoire raisonnée sur l'épizootie de boefs en Piémont et les moyens prompts à y remédier, lu à la Société d'Agriculture de Turin', 1789; both in: *Recueil de mémoires et observations-pratiques sur l'épizootie*, Lyon, Reymann, 1808. On S. L. Mitchill see Courtney Robert Hall, *A scientist in the early republic. Samuel Latham Mitchill 1764–1831*, New York, Russell and Russell, 1962.
38. M. Buniva, *Discours historique sur l'utilité de la vaccination, suivi d'une instruction sur le même objet*, Turin, Imprimerie Départementale, An XII [1804]; M. C. Tirsi, 'L'inizio della vaccinazione in Piemonte e l'epidemia vaiulosa del 1829', *Riv. Storia Med.*, 1974, 18: 64–75.
39. J. Théodoridès, personal communication.
40. M. Buniva, *Observations et expériences sur la maladie épizootique des chats, qui règne depuis quelques années en France, en Allemagne, en Italie et en Angleterre*, Paris, Société de Médecine, 1800.

CHAPTER 6

1. Information on this period in the life of the veterinary profession is to be found in Sir Frederick Smith's authoritative 4 volume work *The early history of veterinary literature and its British development*, London, J. A. Allen and Co., 1976, especially volumes 2 and 3, although some of Sir Frederick's strong and uncompromising opinions must be treated with caution. The background to, and beginnings of, the London Veterinary College were examined by L. P. Pugh in *From farriery to veterinary medicine 1785–1795*, Cambridge, W. Heffer and Sons, 1962. Pugh used the Minute Book of the Odiham Agricultural Society (1783–96), the Minute Book of the Bath and West of England Society (1777–), and the Minute Books of the Veterinary College, London, later the Royal Veterinary College (1790–1851).

2. Miessner, '150 Jahre der Hochschule Hannover', *Dt. tierärzliche Wschr.*, Sondernummer, June 1928, pp. 5–10.

3. Arthur Young's literary activities began with *The Farmer's Letters to the People of England* in 1767, and culminated in his editorship of the journal *Annals of Agriculture*, which first appeared in 1784.

4. W. H. R. Curtler, *A short history of English agriculture*, Oxford, Clarendon Press, 1909; J. D. Chambers and G. E. Mingay, *The agricultural revolution 1750–1880*, London, B. T. Batsford Ltd., 1966; Kenneth Hudson, *The Bath and West*, Bradford-on-Avon, Moonraker Press, 1976.

5. John Brooke, *King George III*, London, Constable, 1972. Two of the sons of George III, the Duke of York and the Prince of Wales, did in fact become early patrons of the London Veterinary College.

6. Arthur Young, *Travels in France during the years 1787, 1788 and 1789,* ed. Constantia Maxwell, Cambridge University Press, 1929, especially pp. 86–7. First published 1792, when Young was full of enthusiasm for the Revolution and the new order in France. Later events served to change his mind.

7. James Clark, *A Treatise on the Prevention of Diseases incidental to Horses*, Edinburgh, W. Smellie, 1788; Preface, p. 4.

8. Quotes from the records of the Odiham Society follow Pugh. The Society's later resolution paraphrased the prose of the original 'Plan' (Pugh's appendix 3), improving not a little on its clarity.

9. H. Hours, 'Les difficultés de l'école de Lyon', in: *La lutte contre les épizooties*, [Paris], 1957, p. 29. The funerary honours were described by Bracy Clark, *Edinb. vet. Rev.*, 1861, 3, p. 134.

10. The remains of Eclipse continued to have links with British veterinary medicine. Having initially passed through a number of private hands, the skeleton is now housed in the National Horseracing Museum at Newmarket. For its history before removal to Newmarket, see Iain Pattison, *John McFadyean*, London and New York, J. A. Allen, 1981, pp. 186–7; also J. W. Barber-Lomax, 'Eclipse and Sainbel', *Vet. Rec.*, 1959, 71: 180.

11. Pugh found a copy of the original plan 'hidden among the books and papers of the Bath and West and Southern Counties Society since 1790' (Pugh, *op. cit.*, Preface). The long struggle for a Royal Charter for the College of Veterinary Surgeons has been described by Iain Pattison, *The British Veterinary Profession 1791–1949*, London, J. A. Allen, 1984.

12. For the rôle of coffee houses in eighteenth century society and science, see W. H. G. Armytage, 'Coffee houses and science', *Br. med. J.*, 1960, ii: 213; and John Keevil, 'Coffee house cures', *J. Hist. Med.*, 1954, 9: 191–5.

13. After this unselfish act the Odiham Society seems to have lost its raison d'être; it faded away and finally went out of existence in 1796.

14. In 1791 Joseph Banks was in the 13th year of his controversial and unprecedentedly long presidency of the Royal Society. Patrick O'Brian, *Joseph Banks: a life*, London, Collins Harvill, 1987.

15. This proved to be a recurrent problem which found a final and legal solution only when the Veterinary Surgeons Bill became law in 1881 (p. 109).

16. W. Hunting, 'Charles Vial de St. Bel', *Vet. Rec.*, 1891–2, 4: 130–3.

17. Sir Frederick Hobday, 'John Hunter – pioneer of veterinary science', *Trans. Hunter. Soc.*, 1937–38, 2: 89–100. Hobday's essay has probably been responsible for several later, exaggerated, references to Hunter's central rôle in the creation of London's veterinary school, at the expense of the Odiham Society and Granville Penn. Pugh, *op. cit.*, sets the record straight.

18. Garry Alder, *Beyond Bokhara. The life of William Moorcroft, Asian explorer and pioneer veterinary surgeon 1767–1825*, London, Century Publishing, 1985.

19. [Youatt] 'The presentation of Mr. Coleman's bust', *The Veterinarian*, 1835, 8: 218–29. On this occasion, the profession's celebration of Coleman's 70th birthday, Youatt behaved with characteristic generosity and restraint in his comments.

20. Not until 1795, and only after personal representations were made to Pitt, did Parliament agree to provide an annual grant. At war with France, under constant threat of invasion, the country was more than ever in need of trained veterinarians to care for cavalry horses.

21. E. Leclainche, *Histoire de la médecine vétérinaire*, Toulouse, Office du Livre, 1936; S. Andersen (ed.), *P. C. Abildgaard (1740–1801)*, Copenhagen, Kandrup, 1985.

22. On this point all authors agree – Smith, Pugh, and Pattison, as well as contemporary sources.

23. William Youatt, *The Dog*, London, Longman, 1851, p. 143. The importance of this volume was recognised in France, where the chapter on rabies appeared, translated from the 1st edition of 1845, with comments, by H. Bouley, in *Rec. méd. vét.*, 1847, 4: 222–50.

24. Youatt, *op. cit.* note 23, p. 144.

25. *ibid.*, p. 145; for Youatt's humanitarian attitude regarding 'humanity to brutes' in its contemporary context, see Keith Thomas, *Man and the natural world*, Penguin Books, 1984, p. 176; C.L., 'Bernard Balzac on hydrophobia', *Br. med. J.*, 1939, ii: 710 [from A. Trillat, *Bull. Acad. Méd.*, 1939, 121: 200–7].

26. J. K. Walton, 'Mad dogs and Englishmen: the conflict over rabies in late Victorian England', *J. Soc. Hist.*, 1979, 13: 219–39.

27. William Youatt, *The Horse with a treatise on draught*, London, Chapman and Hall, 1843.

28. Pattison, *op. cit.* note 11; Clifford H. Eby, 'William Dick, 1793–1866', *Mod. vet. Pract.*, 1958, 39 (10): 36–7.

29. The first issue of the *Journal of the Royal Agricultural Society* was published in 1840. Its editors were able to celebrate both the first anniversary of the English Agricultural Society, in May, 1839, and its Royal Charter granted in March, 1840, which elevated the society to the status of Royal Agricultural Society of England.

30. Ministère de l'Agriculture, *L'Ecole nationale vétérinaire de Toulouse et la profession vétérinaire*, Toulouse, Imprimerie et Librairie Edouard Privat, 1923; M.-R.-M. Clair, *Histoire de la Création de l'Ecole Nationale Vétérinaire de Toulouse*, Toulouse, Ouvrière, 1965.

31. James B. Simonds, *A practical treatise on variola ovina, or, small-pox in sheep*, London, James Ridgway, John Churchill, 1848; *idem, The age of the ox, sheep and pig*, London, W. S. Orr and Co., 1854; *idem*, 'Second report on the prevention of pleuro-pneumonia in cattle by inoculation', *J. Roy. Agr. Soc.*, 1853, 14: 244–73.

32. George Fleming, *Animal Plagues*, London, Chapman and Hall, 1871; I. Katić, *Dansk-russiske veterinaere forbindelser 1796–1976*, Copenhagen, Kandrup, 1982; Ivan Katić (ed.), *Fjernt fra Danmark: breve fra Chr. Engelsens ophold i Rusland til broderen Otto Waldemar Engelsen 1885–1906*, Copenhagen, Kandrup, 1988.

33. J. B. Simonds, 'Report on the cattle plague', *The Veterinarian*, 1859, 32: 414–15.

34. 'Parliamentary intelligence. Extracts from minutes of evidence of the "select committee on the sheep, etc., contagious diseases prevention bill"', *ibid.*, 1857, 30: 466–657, *passim*, and 1858, 31: 33–40; J. B. Simonds, 'Report on the cattle plague, steppe murrain, or rinderpest', *ibid.*, 1858, 31: 92–693, *passim*; John Gamgee, *The Cattle Plague*, London, Robert Hardwicke, 1866; Sherwin A. Hall, 'The cattle plague of 1865', *Med. Hist.*, 1962, 6: 45–58; Pattison, *op. cit.* note 11, p. 55.

35. The story of the Gamgee family has been told by a descendant: Ruth D'Arcy Thompson, *The remarkable Gamgees*, Edinburgh, The Ramsay Head Press, 1974.

36. Gamgee, *op. cit.* note 34, pp. 465–527.

37. George Fleming, *op. cit.* note 32, p. xxxiv.

38. Pattison, *op. cit.* note 11, pp. 92–95, and 182.

39. George Fleming, 'Human and animal variolae: a study in comparative pathology', *Lancet*, 1880, ii: 164–6.

40. Pattison, *John McFadyean, op. cit.* note 10.

41. John McFadyean, 'The suppression of the contagious diseases of animals', *Br. med. J.*, 1894, ii: 758; 'Royal veterinary College. Introductory address by Professor McFadyean', *Lancet*, 1894, ii: 789.

42. [Youatt], 'Medical men and veterinary surgeons', *The Veterinarian*, 1835, 8: 580–5; *cf.* also J. Clark, note 7 above.

43. Thomas D. Brock, *Robert Koch: A Life in Medicine and Bacteriology*, Madison, Wisconsin, Science Tech. Publishers, 1988. In this study Brock demonstrates the gulf between Koch's early, great, contributions, and his later descent into authoritarian arrogance and failure to recognise and admit his mistakes; see also P. Cranefield's review of this volume in *Bull. Hist. Med.*, 1989, 63: 306–8.

44. R. Koch, 'The combating of tuberculosis in the light of the experience that has been gained in the successful combating of other infectious diseases', *Lancet*, 1901, ii: 187–91, p. 189.

45. John McFadyean, 'Tubercle bacilli in cows' milk as a possible source

of tuberculous disease in man', *ibid.*: 268–271; 'The British Congress on Tuberculosis', *ibid*: 302.

46. J. McFadyean, 'The ultravisible viruses', *J. comp. Path. Ther.*, 1908, 21: 58–68; 168–75; 232–42. For the development of the concept of filterable, or ultravisible viruses, see S. S. Hughes, *The virus: a history of the concept*, New York and London, Heinemann Educational Books, 1977; and A. P. Waterson and Lise Wilkinson, *An introduction to the history of virology*, Cambridge University Press, 1978.

CHAPTER 7

1. G. G. Zinke, *Neue Ansichten der Hundswuth, ihrer Ursachen und Folgen, nebst einer sichern Behandlungsart der von tollen Thieren gebissenen Menschen. Für Arzte und Nichtärzte bestimmt*, Jena, C. E. Gabler, 1804.

2. F. Magendie, 'Expérience sur la rage', *J. Physiol. exp. path.*, 1821, 1: 40–6; G. Breschet, 'Note sur des recherches expérimentales relative au mode de transmission de la rage', *C. r. hebd. Séanc. Acad. Sci.*, 1840, 11: 485–501.

3. Ruth Richardson, *Death, Dissection and the Destitute*, London, Routledge and Kegan Paul, 1987, p. 273.

4. Bracy Clark, 'Vial de St. Bel and the early history of the London Veterinary College: being a letter written from Switzerland in 1798', *Edinb. Vet. Rev.*, 1861, 3: 129–37.

5. *The works of Charles Vial de Sainbel, professor of veterinary medicine, to which is prefixed a short account of his life, also the origin of the veterinary college of London*, London, Martin and Bain, 1795; W. Hunting, 'Charles Vial de St. Bel', *Vet. Rec.*, 1891–2, 4: 130–3; L. P. Pugh, *From Farriery to Veterinary Medicine 1785–1795*, Cambridge, W. Heffer and Sons, 1962.

6. Losses among horses in the armies of Europe were well recognised by the middle of the eighteenth century when the elder Lafosse wrote of the 'terrible ravages through this disease' in: E. G. Lafosse, *Traité sur le veritable siège de la morve des chevaux, et les moyens d'y remédier*, Paris, David and Gouchon, 1749, preface.

7. The works of Cagniard-Latour and of Theodor Schwann are discussed in the following chapter (notes 6 and 7); Agostino Bassi, *Del mal del segno calcinaccio o moscardino, malattia che afflige i bachi da seta e sul modo di liberarne le bigatta, e anche le più infestate*, Lodi, 1835 (below, chapter 11); C. G. Ehrenberg, *Die Infusionsthierchen als volkommene Organismen. Ein Blick in das tiefere organische Leben der Natur*, Leipzig, 1838.

8. On Henle see following chapter (note 24).

9. E. Ackerknecht, 'Anticontagionism between 1821 and 1867', *Bull. Hist. Med.*, 1948, 22: 562–693; Margaret Pelling, 'The reality of anti-contagionism theories of epidemic disease in the early nineteenth century', *Soc. social Hist. Med. Bull.*, 1976, no. 17, pp. 5–7.

10. Jean Hameau, *Etude sur les virus* (1836 et 1847), preface par M. Grancher, Paris, Masson, 1895; A. Chabé, 'Jean Hameau (1779–1851)', *Les Biographies médicales*, Paris, J.-B. Baillière et fils, 1934; *Jean Hameau (1779–1851): sa vie et ses oeuvres*, notice publiée par la Commission du Monument Jean Hameau, Bordeaux, G. Gounouilhou, 1899; J. Théodoridès, 'Un précurseur Girondin de la pathologie infectieuse Jean Hameau (1779–1851), *C. r. 104 Congr. nat. Soc. savant.*, Bordeaux, 1979: 81–95.

11. Lorin, 'Observations sur la communication du farcin des chevaux aux hommes', *J. méd. chir. pharm.* 1812, 23: 136–7; Schilling, 'Merkwürdige Krankheits- und Sections-geschichte einer warscheinlich durch Uebertragung eines thierischen Giftes erzeugten Brandrose', *Rust's Magazin für die gesammte Heilkunde*, 1821, 11: 480–503.

12. Andrew Brown, 'Fatal case of glanders in the human subject', *Lond. med. gaz.*, 1829, 4: 134–7.

13. John Elliotson, 'On the glanders in the human subject', *Med. chir. Trans.*, 1830, 16: 171–218.

14. Benjamin Travers, *An inquiry concerning the disturbed state of the vital functions, usually denominated constitutional irritation*, London, Longman, 1826, p. 350; *Lancet*, 1830–31, i: 215.

15. K. H. Hertwig, 'Uebertragung thierischer Ansteckungsstoffe auf Menschen', *Med. Zeitung*, 1834, 3: 215–21; idem., 'Beiträge zur nähern Kenntnis der Wuthkrankheit oder Tollheit der Hunde', *Hufeland's Journal der practischen Arzneykunde und Wundarzneykunst*, 1828, 67: 3–173; for Hertwig's career in the European context, see chapter 11.

16. M. Rooseboom, 'The history of the microscope', *Proc. Roy. Microscop. Soc.*, 1967, 2: 266–93; 'The microscope and medical history', [Report of a symposium at the Science Museum, Sept. 24, 1983], *Med. Sci. hist. Soc. Newsl.*, 1984, no. 3, pp. 4–12; S. Bradbury, *The microscope past and present*, Oxford, Pergamon Press, 1968, pp. 147–57.

17. P. Rayer, *De la morve et du farcin chez l'homme*, Paris, J.-B. Baillière, 1837; G. Breschet et P. Rayer, 'De la morve chez l'homme, des solipèdes et quelques autres mammifères', *C. r. hebd. Séanc. Acad. Sci. Paris*, 1840, 10: 209–23.

18. P. Rayer, 'Sur l'épizootie qui a regné à Paris, dans les derniers mois de 1838 et pendant le 1er semestre de 1839', *Arch. méd. comp.*, 1843, 1: 155–171; idem., 'Fragment d'une étude comparative de la phthisie pulmonaire chez l'homme et chez les animaux', *ibid.*: 189–219.

19. D. W., 'Villemin – pioneer. A chapter in the history of tuberculosis', *Lancet*, 1928, i: 720. *J.-A. Villemin 1827–1892 (centenaire de sa naissance)*, Paris, Masson, 1927.

20. Raoul Caveribert, 'La vie et l'oeuvre de Rayer (1793–1867)', thesis, Paris, M. Vigne, 1932; letter pp. 15–19; J. Théodoridès, 'P.F.O. Rayer (1793–1867), son oeuvre et son influence' in: *Congr. int. Storia Med.* (XXI, Siena, 1968), Rome, E. Cossidente, 1970, pp. 1566–73.

21. 'Pater' [J.H. Tucker], 'Proposed new society for the investigation of

cholera and other epidemic diseases', *Lancet*, 1849, *ii*: 301–302; 'The Foundation of the Society', in: *The Epidemiological Society of London. The Commemoration Volume*, London, 1900, pp. 3–5.

22. Jean Fournier, *Observations sur la nature, les causes,...de la maladie épidémique des chiens*, 1st ed. Dijon, 1764; *idem.*, *Observations sur la nature et le traitement de la fièvre pestilentielle, ou la peste*, Dijon, L. N. Frantin, 1777.

23. W. Cunningham, 'The work of two Scottish medical graduates in the control of wool-sorters' disease', *Scott. Soc. Hist. Med. Rep. Proc.*, 1974–5, in: *Med. Hist.*, 1976, 20: 169–73.

24. [J.] Fournier, *Observations et expériences sur le Charbon Malin*, Dijon, Defay, 1769, pp. 24–30, *passim*.

25. Ph. Chabert, *Ecole Royale Vétérinaire de Paris*, Cap, Imp. Royale, 1778 (signed by Chabert 28 Feb. 1774), 4 pp.

26. [Ph.] Chabert, *Traité du charbon ou anthrax dans les animaux*, Paris, Imp. Royale, 1782. The (largely French) early development of anthrax studies has been discussed in: J. Théodoridès, *Un grand médecin et biologiste Casimir-Joseph Davaine (1812–1882)*, Analecta medico-historica, 4, Oxford, Pergamon Press, 1968, chapter 5, pp. 66–71.

27. e.g. Guillotin, 'Motion pour l'établissement d'un Comité de Santé', Paris, Imp. Nationale, 1790; Guillotin, 'Projet de décret sur l'enseignement de l'art de guérir, présenté au nom du Comité de salubrité', Paris, Imp. Nationale, 1790; Fourcroy, 'Rapport et projet de décret sur l'établissement d'une Ecole Centrale de Santé, à Paris', Imprimés par ordre de la Convention Nationale, 1794. Original French decrees discussed by Caroline Hannaway in series of lectures at Wellcome Institute, London, May–June 1989.

28. Work referred to in: 'Compte rendu des travaux de l'Ecole Vétérinaire d'Alfort, pour l'année 1823', *Rec. Méd. vét.*, 1824, 1: 144–50.

29. Carl von Linné, *Amoenitates Academicae*, vol. 3, p. 322; English translation in: *A general system of nature*, by Sir Chas. Linné, London, Lackington, Allen and Co., 1806, vol. 4, pp. 56–7.

30. Théodoridès, *op. cit.*, gives an objective account in his Davaine biography; not so Reiner Müller, whose sympathy for his German compatriots leads him to somewhat exaggerated and unbalanced statements, Reiner Müller, '80 Jahre Seuchenbakteriologie. Die Seuchenbackteriologen vor Robert Koch: Pollender 1849, Brauell 1856, Delafond 1856, Davaine 1863', *Zentbl. Bakt. ParasitKde*, Abt. 1, Orig. 1929, 115: 1–17.

31. P. Rayer, 'Inoculation du sang de rate', *C. r. Séanc. Soc. Biol.*, 1850, 2: 141–4; E. Lagrange, *Robert Koch. Sa vie et son oeuvre*, Paris, Legrand, and Bruxelles, Edit. Universelles, 1938, p. 12.

32. A. Pollender, 'Mikroskopische und mikrochemische Untersuchung des Milzbrandblutes sowie über Wesen und Kur des Milzbrandes', *Casper's Vierteljahrschr. gerichtl. öffentl. Med.*, 1855, 8: 103–14.

33. Fr. A. Brauell, 'Versuche und Untersuchungen betreffend den Milz-brand des Menschen und der Thiere', *Arch. Path. Anat. Physiol.*, 1857,

11: 132–44; *idem*, 'Weitere Mittheilungen über Milzbrand und Milzbrandblut', *ibid.*, 1858, 14: 432–66. For details of Brauell's career, and the state of veterinary pathology in the Russian empire see Leon Z. Saunders, *Veterinary Pathology in Russia, 1860–1930*, Ithaca and London, Cornell University Press, 1980.

34. Henri M.-O. Delafond, 'Communication sur la maladie regnante', *Bull. Soc. Centr. Vét.* in: *Rec. Méd. Vét.*, 1860, 37: 726–48.

35. Louis Pasteur, 'Expériences et vues nouvelles sur la nature des fermentations', *C. r. hebd. Séanc. Acad. Sci. Paris*, 1861, 52: 1260–4; C. Davaine, 'Recherches sur les infusoires du sang dans la maladie connue sous le nom de sang de rate', *ibid.*, 1863, 57: 220–3; *idem*, 'Nouvelles recherches sur les infusoires du sang dans la maladie connue sous le nom de sang de rate', *ibid.*: 351–3; 386–7; *idem*, 'Nouvelles recherches sur la maladie du sang de rate, considerée principalement au point de vue de sa nature', *Mém. Soc. Biol.*, 3ᵉ sér., 1863, 5: 193–202. Théodoridès, *op. cit.* note 26, has discussed all Davaine's publications in depth.

36. C. Davaine, 'Nouvelles recherches sur la nature de la maladie charbonneuse connue sous le nom de "sang de rate"', *C. r. hebd. Séanc. Acad. Sci. Paris*, 1864, 59: 393–6.

37. Davaine et Raimbert, 'Sur la présence de bactéridies dans la pustule maligne chez l'homme', *ibid*: 429–31.

38. C. Davaine, 'Sur la nature des maladies charbonneuses', *Arch. gen. Méd.*, 6ᵉ sér., 1868, 2: 144–8; *idem.*, 'Etudes sur la contagion du charbon chez les animaux domestiques', *Bull. Acad. Imp. Méd.*, 1870, 35: 215–35.

39. J. Théodoridès, *op. cit.* note 26, pp. 20–1.

CHAPTER 8

1. The early development of and thinking on pyaemia and septicaemia has been discussed at some length by William Bulloch in *The history of bacteriology*, Oxford University Press, 1838, chapters 5 and 6. R. G. Mayne in *Expository Lexicon*, London, John Churchill, 1860, defined 'pyohaemia, also spelt pyaemia' but did not include septicaemia; Tyndall, in various editions of *Fragments of Science* throughout the 1870s used both terms.

2. B. Gaspard, 'Second mémoire physiologique et médicale sur les maladies putrides', *J. physiol. exp. path.* 1824, 4: 1–69, p. 3.

3. *idem.*, 'Mémoire physiologique sur les maladies purulentes et putrides, sur la vaccine, etc.', *ibid.*, 1822, 2: 1–45, p. 37.

4. [F.] Magendie, 'Remarques sur la notice précédente, avec quelques expériences sur les effets des substances en putréfaction', *ibid.*, 1823, 3: 81–8, p. 83.

5. *ibid.*, pp. 84–5.

6. Th. Schwann, 'Vorläufige Mittheilung betreffend Versuche über die Weingärung und Fäulniss', *Ann. Physik u. Chemie* 1837, 41: 184–93.

7. [C.] Cagniard-Latour, 'Mémoire sur la fermentation vineuse', *C. r. hebd. Séanc. Acad. Sci. Paris* 1837, 4: 905–6; details in: *L'Institut* (*J. gén. Soc. trav. sci. France*), 1836, nos 164–7, 185; 1837, no. 199, *passim*.

8. P.-A. Piorry, *Traité de Diagnostic et de Séméiologie*, Bruxelles, Société typographique Belge, A. D. Wahlen Co., 1837, pp. 117–18; *Exposé analytique...de P. A. Piorry*, Paris, J.-B. Baillière, 1856, pp. 35–40.

9. Rudolf Virchow, 'Thrombose und Embolie. Gefässentzündung und septische Infektion', in: *Gesammelte Abhandlungen zur Wissenschaftlichen Medicin*, Frankfurt a.M., von Meidinger, 1856, part 4, pp. 219–732; experiments with dogs, pp. 264–85.

10. C. E. Sédillot, *De l'infection purulente ou pyoémie*, Paris, 1849.

11. P. L. Panum, 'Bidrag til Laeren om den saakaldte putride eller septiske Infection', *Bibliothek for Laeger*, 1856, 8: 253–85. For an evaluation of Panum's scientific contributions see A. Gjedde, *Peter Ludvig Panums videnskabelige indsats*, Copenhagen, Costers, 1971.

12. B. Gaspard, note 2 above, exp. 42, p. 43.

13. [C.] Davaine, 'Recherches sur quelques questions relatives à la septicémie', *Bull. Acad. Méd.* 1872, 1, 2 ser.: 907–29; 976–96, p. 989.

14. O. M. Röthlin, *Edwin Klebs 1834–1913*, Zürich, Juris-Verlag, 1962.

15. E. Klebs, 'Ueber Diphtherie', 1883, reprinted *Bull. Hist. Med.*, 1940, 8: 509–22; F. Loeffler, 'Untersuchungen über die Bedeutung der Mikroorganismen für die Entstehung der Diphtherie beim Menschen, bei der Taube und beim Kalbe' (Dec. 1883), *Mitt. Ksl. Gesdh. amt.* 1884, 2: 421–99.

16. E. T. Tiegel, 'Die Ursache des Milzbrandes', *Korresp. Bl. schweizer Arzte*, 1871, 1: 275–80.

17. E. Klebs, *Beiträge zur pathologischen Anatomie der Schusswunden*, Leipzig, Vogel, 1872.

18. Tiegel, op. cit. note 16.

19. C. J. Eberth, *Zur Kenntniss der bacteritischen Mycose*, Leipzig, 1872.

20. L. Pasteur, 'Nouvelles expériences relatives aux générations dites spontanées', *C. r. hebd. Séanc. Acad. Sci. Paris*, 1860, 51: 348–52; *idem*, 'Mémoire sur les corpuscles organisés qui existent dans l'atmosphère. Examen de la doctrine des générations spontanées', *Ann. Sci. nat.*, 1861, 16: 5–98.

21. F. Cohn, *Ueber Bacterien, die kleinsten lebenden Wesen*, Berlin, Carl Habel, 1872.

22. H. Mochmann and W. Köhler, *Meilensteine der Bakteriologie*, Jena, Gustav Fischer Verlag, 1984, pp. 148–50.

23. Thomas D. Brock, *Robert Koch. A life in medicine and bacteriology*, Madison, WI, Science Tech. Publishers, 1988; see chapter 6, 'Koch and Cohn'.

24. J. Henle, 'Von den Miasmen und Contagien und von den miasmatisch-contagiösen Krankheiten', in: *Pathologische Untersuchungen*, Berlin, A. Hirschwald, 1840, pp. 1–82, p. 15; for English translation see George Rosen, 'Jacob Henle "On miasmata and contagia"', *Bull. Hist. Med.*, 1938, 6 (2): 907–83, p. 923.

25. [—], 'Dr. Henle's Pathological Researches', *Brit. For. Med. Rev.*, 1840, 9: 398–410, p. 398.
26. J. Henle, op. cit. note 24, pp. 16–17; Rosen, p. 925.
27. *ibid.*, Henle, pp. 76–80; Rosen, pp. 977–81.
28. H. M. Woodcock, 'The nature of viruses', *Lancet*, 1931, ii: 936.
29. F. H. Garrison, *History of medicine*, Philadelphia and London, W. B. Saunders Co., 1929, p. 458.
30. *Cf.* chapters 3 and 4 above.
31. Sir Henry Holland, 'The hypothesis of insect life as a cause of disease?', *in: Medical notes and reflections*, London, Longman, 1839, pp. 560–89.
32. Neale's exploits are recorded in the *DNB*.
33. Adam Neale, *Researches to establish the truth of the Linnaean doctrine of animate contagions...*, London, Longman, 1831, p. 5.
34. *ibid.*, p. 12.
35. *ibid.*, pp. 151–2.
36. *ibid.*, p. 242.
37. *ibid.*, p. 42.
38. William Coleman, *Yellow fever in the North. The Methods of Early Epidemiology*, University of Wisconsin Press, 1987; for Chervin see chapter 2, esp. pp. 26–30.
39. W. Fergusson, *Notes and recollections of a professional life*, London, Longman, 1846, p. 100.
40. *idem.*, 'Notes and observations upon the contagion of typhous fever, and contagion generally', reprinted in the above, pp. 162–84; see p. 163.
41. W. Fergusson, 'Letters upon the cholera morbus', London, Highley, 1832, p. 7.
42. *Cf.*, Coleman, note 38 above.
43. 'First annual report of the Registrar-General of Births, Deaths, and Marriages in England', *Brit. For. Med. Rev.* 1840, 9: p. 344.
44. William Farr, *Vital Statistics*, ed. M. Susser and A. Adelstein, Metuchen, NJ, Scarecrow Press Inc., 1975, p. 333.
45. *ibid.*, pp. 317–18.
46. *ibid.*, pp. 328–30.
47. J. Brownlee, 'Historical note on Farr's theory of the epidemic', *Brit. med. J.*, 1915, ii: 250–2.
48. William Budd, 'Variola ovina, sheep's small-pox; or the laws of contagious epidemics illustrated by an experimental type', *ibid.*, 1863, ii: 141–50, p. 141.
49. B. W. Richardson, 'The present position and prospects of epidemiological science', *ibid.*, pp. 597–601.
50. W. Budd, 'Investigation of epidemic and epizootic diseases', *ibid.*, 1864, ii: 354–7.
51. Budd, *op. cit.* note 48, p. 142.
52. *ibid.*, f.n.
53. *ibid.*, p. 144; *cf.* also chapter 7 above.
54. W. Cunningham, 'The work of two Scottish medical graduates in the control of woolsorters' disease', *Med. Hist.*, 1976, 20: 169–73.

55. John M. Eyler, 'William Farr on the cholera: the sanitarian's disease theory and the statistician's method', *J. Hist. Med.*, 1973, **28**: 79–100, see pp. 94–5.

56. J. Burdon Sanderson, 'On the cattle plague in its epidemiological aspects', *Trans. Epid. Soc.*, 1967, **2** (2): 463–70.

57. *ibid.*, p. 465.

CHAPTER 9

1. J. Théodoridès, *Un grand médecin et biologiste: Casimir-Joseph Davaine (1812–1882)*, Analecta medico-historica, 4, Oxford, Pergamon Press, 1968.

2. Raoul Caveribert, 'La vie et l'oeuvre de Rayer (1793–1867)', thesis, Paris, M. Vigne, 1931, pp. 15–19.

3. This was in sharp contrast to the English system, where in the schools, the public schools and the universities, classics, arts and humanities in general enjoyed a degree of preferential treatment which left scientific subjects in a position little short of disregard. This disadvantaged educational position discouraged a majority of the abler students from entering the field of science, and consequently was responsible for certain relative weaknesses in British science in the nineteenth and well into the twentieth century. Thomas Arnold of Rugby saw no place for science in liberal education. See M. McCrum, *Thomas Arnold, a reassessment*, Oxford University Press, 1990.

4. *Histoire et Mémoires de la Société Royale de Médecine*, Paris, Didot jeune, 1779, pp. 1–2; Caroline C. Hannaway, 'The Société Royale de Médecine and epidemics in the Ancien Régime', *Bull. Hist. Med.*, 1972, **46**: 257–73.

5. The first and only volume of Rayer's *Arch. méd. comp.* was published in 1843; *cf.* chapter 7.

6. See chapter 5 above.

7. Additional specialised training for the more promising students was available at the Alfort school after 1815, *cf.* p. 97.

8. J. Bost and C. de Lourdes Branco, 'Chauveau Docteur en Médecine', *Bull. Soc. Sci. Vét. Lyon*, No. 4, 1969: 319–22.

9. F.-X. Lesbre, 'Notice sur la vie et les travaux de J.-B.-A. Chauveau', Lyon, A. Rey, 1917; F. Arloing, 'Inauguration du Monument J.-B.-A. Chauveau à l'Ecole vétérinaire de Lyon', Toulouse, J. Bonnet, 1927; G. Ramon, 'Hommage à Jean-Baptiste Auguste Chauveau', *Bicentenaire de l'Ecole Nationale Vétérinaire de Lyon* (25–26 mai 1962); G. Legee, 'La chaire de pathologie comparée du Muséum national d'Histoire naturelle et l'oeuvre de A. Chauveau (1827–1917)', *Hist. Nat.*, 1976, no. 8: 53–88.

10. C. Leblanc, 'Discours au nom de l'Academie de Médecine', *in: Henri Bouley 17 Mai 1814–30 Novembre 1885*, volume of obituary notices, '*éloges*', etc., Paris, Renov et Maulde, 1885, pp. 23–7.

11. Chapter 7, p. 126, and note 28.

12. *Cf.* chapters 5 and 6.

13. See the extensive volumes of *Bull. Acad. Imp. Méd.* for the 1860s, e.g. vol. **27**, 1861–2, pp. 835–61; 877–81; 883–909; vol. **28**, 1862–3, pp. 553–66; 576–85; vol. **29**, 1863–4, pp. 140–66; 218–43; 1039–58; 1137–56.

14. A. Goubaux, 'Discours au nom des Ecoles Vétérinaires de France', *in:* *Henri Bouley...*, note 10 above, pp. 33–9.

15. e.g. 'Discussion sur la pustule maligne', *Bull. Acad. Imp. Méd.*, 1863–4, **29**: 1137–56, esp. pp. 1137–40.

16. Educated at Montpellier, Bousquet settled in Paris and was for a number of years the capital's authority on vaccination. He founded the *Rev. méd.* in 1820, and also edited the *Bull. Acad. Méd.* between 1836 and 1850.

17. [J.-B.-E.] Bousquet, 'De l'origine de la vaccine sur le cheval', *Bull. Acad. Méd.*, 1861–2, **27**: 835–50.

18. *Op. cit.* note 15, see p. 1141.

19. 'Discussion sur l'origine de la vaccine', *Bull. Acad. Méd.*, 1861–2, **27**: 883–909; *ibid.*, 1863–4, **29**: 218–43. Cf. D. Baxby, *Jenner's smallpox vaccine*, London, Heinemann Educational Books, 1981.

20. See Goubaud, note 14 above, pp. 35–6.

21. J. S. Oxford, 'What is the true nature of epidemic influenza virus and how do new epidemic viruses spread?', *Epidem. Inf.*, 1987, **99**: 1–3; Sir F. Hoyle and Chandra Wickramasinghe, 'Does epidemic disease come from space?', *New Scient.*, 1977, **76**: 402–4.

22. See, for example, J. Bost, 'A propos du registre du laboratoire de Chauveau (Mars–Novembre 1861): l'histoire des premiers enregistrements cardiographiques', *Hist. Sci. Méd.*, 1974, **8**: 595–626.

23. [A.] Chauveau, 'Recherches experimentales de la Société des sciences médicales de Lyon sur les relations qui existent entre la variole et la vaccine', *Bull. Acad. Méd.*, 1865, **30**: 808–16, see pp. 814–15.

24. See William B. Dean, 'Walter Reed and the ordeal of yellow fever experiments', *Bull. Hist. Med.*, 1977, **51**: 75–92.

25. A. Chauveau, 'Des prétendues émanations virulentes volatiles et de l'état sous lequel les virus sont jetés dans l'atmosphère par les sujets atteints de maladies contagieuses', *C. r. hebd. Séanc. Acad. Sci.*, 1871, **73**: 116–18; *idem*, 'La science et la législation dans leur rapports avec la police sanitaire du typhus épizootique ou peste bovine, en France', *Rev. sci.*, 1871, I, 2 ser.: 77–87. The earlier part of this volume, published in the wake of the siege of Paris and the armistice, also contains among other advice on public health during the siege, 'La peste bovine et la consommation des bêtes atteintes', by Bouley, *ibid.*, pp. 38–9.

26. For example his analysis of Toussaint's contributions to anthrax vaccination, *C. r. hebd. Séanc. Acad. Sci.*, 1882, **94**: 1694–8.

27. See René Vallery-Radot, *The Life of Pasteur*, London, Constable, 1919.

28. See chapter 2, 'The methodological background', in: A. P. Waterson

and Lise Wilkinson, *An introduction to the history of virology*, Cambridge University Press, 1978.

29. J. Burdon Sanderson, Introductory report on 'The intimate pathology of contagion'. In Appendix to *12th Annual report of the Medical Committee of the Privy Council*, London, Eyre and Spottiswoode, 1869, p. 233.

30. J. P. Morat, 'H. Toussaint; son oeuvre', *Bull. Lyon Méd.*, 1890, **65**: 55–65; G. Ramon, 'Contribution des vétérinaires à la recherche scientifique en biologie et en médecine expérimentale et comparée', *Bicentenaire de l'Ecole Nationale Vétérinaire de Lyon (25–26 mai 1962)*.

31. J. Théodoridès, 'Quelques grands précurseurs de Pasteur', *Hist. Sci. méd.*, 1972, **6**: 336–43, see especially p. 339.

32. H. Toussaint, 'De l'immunité pour le charbon acquise à la suite d'inoculations préventives', *C. r. hebd. Séanc. Acad. Sci.*, 1880, **91**: 135–7; see also *idem*, 'Identité de la septicémie expérimentale aiguë et du choléra des poules', *ibid.*: 301–4; and L. Pasteur, 'Sur les maladies virulentes et en particulier sur la maladie appelée vulgairement choléra des poules', *ibid.*, 1880, **90**: 239–48; *idem*, 'De l'attenuation du virus du choléra des poules', *ibid.*, 1880, **91**: 673–80.

33. L. Pasteur, [C.] Chamberland and [E.] Roux, 'Le vaccin du charbon', *ibid.*, 1881, **92**: 666–8; *idem*, 'Compte rendu sommaire des expériences faites à Pouilly-le-Fort près Melun, sur la vaccination charbonneuse', *ibid.*: 1378–83.

34. Ch. Chamberland and E. Roux, 'Sur l'atténuation de la virulence de la bactéridie charbonneuse, sous l'influence des substances antiseptiques', *ibid.*, 1883, **96**: 1088–91; *idem*, 'Sur l'atténuation de la bactéridie charbonneuse et de ses germes sous l'influence des substances antiseptiques', *ibid.*: 1410–12.

35. J. B. A. Chauveau, 'Etude expérimentale des conditions qui permettent de rendre usuel l'emploi de la méthode de M. Toussaint pour atténuer le virus charbonneux', *ibid.*, 1882, **94**: 1694–8; *idem*, 'De l'atténuation directe et rapide des cultures virulentes par l'atténuation par la chaleur', *ibid.*, 1883, **96**: 553–7.

36. Robin Yves, *Vie et oeuvre de P. V. Galtier (1846–1908)*, Lyon, Camille Annequin, 1957.

37. L'Ecole Nationale Vétérinaire de Lyon, *Inauguration du Buste du Professeur Galtier*, Lyon. A. Rey, 1913.

38. *Op. cit.* note 36, pp. 41–68.

39. V. Galtier, 'Etudes sur la rage', *Ann. Méd. Vét.*, 1879, **28**: 627–39, pp. 627–8.

40. Zinke had used rabbits, among a number of other animal species, for transmission experiments as early as 1804, *cf.* chapter 7.

41. *Op. cit.* note 39, p. 637; also *C. r. hebd. Séanc. Acad. Sci.*, 1879, **89**: 444–6.

42. See J. Théodoridès, 'Un precurseur de Pasteur: Pierre-Victor Galtier (1846–1908)', *Arch. Intern. Claude Bernard*, 1972, no. 21, pp. 1–5.

43. P. V. Galtier, 'Les injections de virus rabique dans le torrent circulatoire

ne provoquent pas l'éclosison de la rage et semblent conférer l'immunité. La rage peut être transmise par l'injection de virus rabique', *C. r. hebd. Séanc. Acad. Sci.*, 1881, **93**: 284–5.

44. Jean Théodoridès, *Histoire de la rage*, Paris, Masson, 1986.

45. L. Pasteur (with C. Chamberland and Roux), 'Sur une maladie nouvelle, provoquée par la salive d'un enfant mort de la rage', *C. r. hebd. Séanc. Acad. Sci.*, 1881, **92**: 159–65.

46. For Roux's life and contributions see E. Lagrange, *Monsieur Roux*, Brussels, Goemaere, 1954; and also *Obit. Not. Fell. R. Soc. Lond.*, 1932–5, **1**: 197–204.

47. See A. Delaunay, *L'Institut Pasteur des origines à aujourd'hui*, Paris, Editions France-Empire, 1962.

48. Although he had recovered well from a stroke, Pasteur's health was failing in his later years.

49. Roux later claimed that the first Chamberland filter had been developed from the stem of a clay pipe, see *Bull. Assoc. anc. élèves Inst. Past.*, 1971, no. 49.

50. Even then, inclusion of the rabies virus in this category with any certainty continued to present difficulties, *cf. Med. Hist.*, 1977, **21**, pp. 23–7.

51. See Thomas D. Brock, *Robert Koch*, Madison, WI, Science Tech. Publ., 1988, pp. 153–5, 'The French Expedition'.

52. See Théodoridès, *op. cit.* note 44, chapter 7.

53. *ibid.*

54. J. Théodoridès, 'Pasteur and rabies: the British connection', *J. Roy. Soc. Med.*, 1989, **82**: 488–90.

55. Ph. Descourt, 'Rensignements préliminaires sur "une histoire méconnue": l'invention des vaccins modernes aux XIXe siècle, *Arch. int. Cl. Bernard*, 1974, no. 5: 165–84; Théodoridès, *op. cit.* note 31.

56. See Stephen Paget, *Sir Victor Horsley*, London, Constable, 1919, p. 75.

57. A. Delaunay, *op. cit.* note 47; and René Vallery-Radot, *op. cit.* note 27.

58. See E. Sergent, 'L'Institut Pasteur du Maroc à Casablanca', *Annls Inst. Past. Paris*, 1933, **51**: 5–13; also obituaries of Albert Calmette, *ibid.*, pp. 589–94, and of Paul Remlinger, *ibid.*, 1965, **108**: 689–94.

59. A. P. Waterson and Lise Wilkinson, *An introduction to the history of virology*, Cambridge University Press, 1978, pp. 32–3.

60. See section on vaccine production and tissue culture in Joan Crick and Arthur King, 'Culture of rabies virus *in vitro*', in: *Rabies*, J. B. Campbell and K. M. Charlton (eds), Kluwer Acad. Publ., 1988, pp. 59–60; and 'Rabies vaccines and immunity to rabies', in: Colin Kaplan, G. S. Turner, and D. A. Warrell, *Rabies: the facts*, Oxford University Press, 1986, pp. 8–20.

61. Like a number of French institutions and universities, the institute today produces its own wine, a Beaujolais: *Chateau des Ravatys*, Cote-de-Brouilly, Domaine de l'Institut Pasteur. I am indebted to Dr Alex Edelman for providing a sample.

CHAPTER 10

1. See J. Burdon Sanderson, 'Introductory Report on "the intimate pathology of contagion", in Appendix to *12th Annual Report of the Medical Officer of the Privy Council*, London, Eyre and Spottiswoode, 1869, pp. 229–56, esp. pp. 232–4.

2. '...extract from the Will of the late Thomas Brown, Esq....', *University of London, Senate Minutes*, Vol. 3, 1850–4, pp. 6–10; the history of the Brown Animal Sanatory Institution has been told by Sir Graham Wilson in 4 articles in the *Journal of Hygiene, Camb.*: 1979, 82: 155–176; 337–52; 501–21; and 83: 171–97.

3. R. B. McDowell and D. A. Webb, *Trinity College Dublin 1592–1952*, Cambridge University Press, 1982, p. 232; W. Macneile Dixon, *Trinity College, Dublin*, London, F. E. Robinson & Co., 1902, p. 196.

4. *Un. Lond. Sen. Mins*, vol. 4, 1855–8, pp. 16–19 (13 May, 1857); *Public Record Office*, Chancery Reports, C:33:959.

5. *Un. Lond. Sen. Mins*, vol. 6, 1863–6, 19 April, 1865 and 18 October, 1865.

6. *Un. Lond. Sen. Mins*, vol. 7, 1867–70: 195, p. 89.

7. 'The Brown Bequest', *Lancet*, 1870, ii: 686.

8. William Bulloch feelingly described the Brown and its surroundings in its days of decline, W.B., 'In Memoriam Emanuel Klein 1844–1925', *J. Path. Bact.*, 1925, 28: 684–97, p. 686. Buildings and arrangements at the time of the institution's opening were described in the *Lancet*, 1871, ii: 654 (4 November).

9. *Br. med. J.*, 1871, i: 16. Italics in quotes are mine.

10. *Lancet*, 1872, ii: 931.

11. Evidence given before Royal Commissions on Vivisection by staff and visitors working at the Brown Institution can be found in all Commission Reports from the 1870s and into the twentieth century. See also, e.g., 'Debate on vivisection at the meeting of Convocation of the University of London', *Lancet*, 1874, i: 708–709; and 'Vivisection at the Brown Institute', *Vet. Rec.*, 1903 (4 April), p. 631.

12. *Lancet*, 1871, ii: 654.

13. One such exception was Burdon Sanderson's paper 'On the Cattle-Plague in its Epidemiological Aspects', *Trans. Epid. Soc.*, Vol. II, part II, Session 1865–6, pp. 463–70.

14. Before securing the Brown appointment for himself, Burdon Sanderson had pursued his experimental interests in a private laboratory in Howland Street; see Sir Edward Sharpey-Schafer, *History of the Physiological Society, 1876–1926*, Cambridge University Press, 1927.

15. For examples of the arguments taking place on the subject towards the end of the century, see 'The reconstitution of the University of London', *Lancet*, 1891, i: 1110–11, and *Br. med. J.*, 1900, ii: 1266. For further details, see Negley Harte, *The University of London 1836–1986*, London, Athlone Press, 1986.

16. Total numbers of animals treated were given pride of place in the

excerpts from the Annual Reports published in the *Lancet* and the *British Medical Journal*. Horses figured prominently, followed by dogs; other species rarely counted for more than 10 per cent of the total.

17. J. Burdon Sanderson, 'Criticisms of Dr. Chauveau of Lyons on the discussion at the Pathological Society on pyaemia', *Br. med. J.*, 1872, ii: 459–60.

18. [—], 'William Duguid', *Vet. Rec.*, 1900 (20 January), p. 406.

19. Answer to question no. 3620 at Royal Commission on Vivisection 1874; see W. Bulloch, *op cit.* note 8, p. 686. Klein first came to London, and was noticed by Burdon Sanderson, when he was sent by Salomon Stricker (1834–98) to negotiate terms for a translation of Stricker's *Handbuch von den Geweben des Menschen und der Thiere*.

20. J. Burdon Sanderson, 'Preliminary Report on the Brown Institution', *J. Roy. Agric. Soc.*, 2nd ser., 1876, 12: 542–4; *idem*, 'Concluding report on the experiments of the Brown Institution on pleuro-pneumonia', *ibid.*, 1879, 15: 157–73; W. Duguid, 'Report on the health of animals of the farm in 1877', *ibid.*, 1878, 14: 233–8.

21. Burdon Sanderson, 1879, *op. cit.* note 20.

22. Ed., 'Foreign view of the Norwich vivisection trial', *Lancet*, 1874, ii: 916.

23. By 1880, attitudes were beginning to change. George Fleming, ever ready to urge the medical profession to include the study of animal diseases and comparative pathology in the curriculum of medical schools (*Lancet*, 1880, i: 164), addressed the British Medical Association on the importance of destruction of tuberculous cattle and the danger of using their flesh and milk. *Br. med. J.*, 1880, ii: 473.

24. See I. Pattison, *John McFadyean*, London, J. A. Allen, 1981; and *idem*, *The British Veterinary Profession 1791–1948*, London, J. A. Allen, 1984.

25. *Univ. Lond. Sen. Mins*, 1907–8, vol. 27, 1675–2378, *passim*; also Univ. Lond. Archives, MSS CF 1/8/15.

26. All published in *J. Roy. Agric. Soc.*, *cf.* note 20 above.

27. See 'The Brown Institution', *Br. med. J.*, 1880, i: 568.

28. [J.] Burdon Sanderson, 'Report on experiments on anthrax conducted at the Brown Institution, February 18th to June 30th, 1878', *J. Roy. Agric. Soc.*, 1880, 16: 267–73.

29. Lady Burdon Sanderson, *Sir John Burdon Sanderson. A Memoir*, Oxford, Clarendon Press, 1991, p. 149.

30. Ed., 'The Brown Institute', *Br. med. J.*, 1878, ii: 848. See also W. D. Tigertt, 'William Smith Greenfield...', *J. Hyg., Camb.*, 1980, 85: 415–20.

31. W. S. Greenfield, 'Lectures on some recent investigations into the pathology of infectious and contagious diseases', *Lancet*, 1880, i: 865–7; *idem.*, 'Report on an experimental investigation on anthrax and allied diseases, made at the Brown Institution', *J. Roy. Agric. Soc.*, 1881, 42: 30–44; *idem.*, 'Preliminary note on some points in the pathology of anthrax, with especial reference to the modification of the

properties of the *Bacillus anthracis* by cultivation, and to the protective influence of inoculation with a modified virus', *Proc. R. Soc. Lond.*, 1880, 30: 557.

32. A. Toussaint, 'De l'immunité pour le charbon, acquise à la suite d'inoculations préventives', *C. r. hebd. Séanc. Acad. Sci.*, 1880, 91: 135–37; *cf.* preceding chapter.

33. Pasteur, Chamberland and Roux, 'Le vaccin du charbon', 1881, 92: 662–5; 666–8; *cf.* also notes 33 and 34, chapter 9.

34. 'Paris. From our own correspondent', *Lancet*, 1880, ii: 750–2; 'M. Pasteur's inoculations', *ibid.*: 782–3.

35. W. S. Greenfield, 'Inaugural address on pathology, past and present', *Lancet*, 1881, ii: 738–41; 781–6, see p. 784.

36. J. Simon, 'Inaugural address', *Trans. Intern. Med. Congr.* (ed. W. MacCormac), London, J. W. Klockmann, 1881, vol. 4, pp. 409–16.

37. 'The Brown Institution', *Lancet*, 1871, ii: 654.

38. *Lancet*, 1873, ii: 238; *cf.* also Tilly Tansey, 'Charles Sherrington and the Brown Animal Sanatory Institution', *St. Thomas's Hospital Gazette*, Spring, 1986.

39. Stephen Paget, *Sir Victor Horsley. A Study of his life and work*, London, Constable and Co., 1919, p. 51.

40. 'The Brown Institution', *Lancet*, 1884, i: 587–8.

41. C.S.S. 'C. S. Roy 1854–1897', *Obit. Not. Fell, R. Soc. Lond.* 1898–1904, part ii, pp. 131–6; Raymond Williamson, 'A photograph of Sir Charles Sherrington and Professor Charles Smart Roy and three letters by Sir Charles Sherrington', *Med. Hist.*, 1959, 3: 78–81.

42. The standard biography of Victor Horsley is Stephen Paget's, *op. cit.* note 39.

43. Ed., 'Rabies and hydrophobia', *Lancet*, 1885, ii: 1054–5. Already ten years earlier, Burdon Sanderson had felt the need to address the public through the pages of the *Times*, to warn of the dangers and outline the symptoms of rabies in the dog, see 'Mad dogs', *Lancet*, 1874, i: 691–2.

44. Paget, *op. cit.* note 39, pp. 74–5.

45. 'The Report of the Commission on Hydrophobia', *Br. med. J.* 1887, ii: 27–9.

46. J. S. Bristowe and Victor Horsley, 'A case of paralytic rabies in man, with remarks', *Trans. Clin. Soc. London*, 1889, 22: 38–47.

47. The striking reduction in cases of rabies in Lancashire following the introduction of the Muzzling Order in November, 1887, was noted by Horsley in 1888 in a report which also warned of the threat of rabies from Surrey working its way into London at a time when deer in Richmond Park were dying from rabies: 'Report of the Brown Institution for 1887', *Br. med. J.*, 1888, i: 815. See also Annual Reports in Senate Minutes for 1887–8.

48. See Paget, *op. cit.* note 39, p. 75, and pp. 87–9.

49. See Annual Reports in Senate Minutes for 1887–91. Horsley also collaborated with Charles E. Beevor in experiments on bonnet monkeys

and orang-outangs, see e.g., *Phil. Trans. R. Soc. Lond.*, 1890, 181:
49–86; 129–58. Also published by the Royal Society were Horsley's
studies on myxoedema and the thyroid gland.

50. See Tilly Tansey, *op. cit.* note 38.

51. 'Obituary. Sir John Rose Bradford', *Br. med. J.*, 1935, i: 805–7; also
Lancet, 1935, i: 906–8; 'Obituary. Thomas Gregor Brodie', *Br. med. J.*,
1916, ii: 342.

52. Paul Fildes, 'Frederick William Twort 1877–1950', *Obit. Not. Fell. R.
Soc. Lond.* 1950–1, 7: 505–17.

53. F. W. Twort, 'The influence of glucosides on the growth of acid-fast
bacilli, with a new method of isolating human tubercle bacilli...', *Proc.
Roy. Soc. Lond. B*, 1907, 79: 329–36; *idem.*, 'A method for isolating
and growing the lepra bacillus of man', *ibid.*, 1910, 83: 156–8.

54. F. W. Twort and G. L. Y. Ingram, 'A method for isolating and
cultivating the *Mycobacterium enteritidis chronicae pseudotubercu-
losae bovis*, Jöhne,...', *ibid.*, 1911–12, 84: 517–42; *idem.*, *A monograph
on Johne's disease*, London, Baillière, Tindall and Cox, 1913.

55. F. W. Twort, 'An investigation on the nature of ultra-microscopic
viruses', *Lancet*, 1915, ii: 1241–3; F. d'Herelle, *Le bacteriophage*, Paris,
Masson, 1921; *idem.*, *The bacteriophage*, Baltimore, Williams and
Wilkins Co., 1922.

56. See A. P. Waterson and L. Wilkinson, *An introduction to the history of
virology*, Cambridge University Press, 1978, chapters 8 and 9, 'The
bacterial viruses'; also J. Cairns, G. S. Stent, and J. D. Watson (eds),
Phage and the origins of molecular biology, Cold Spring Harbor
Laboratory of Quantitative Biology, 1966.

57. See G. Wilson, 'The Brown Animal Sanatory Institution', *J. Hyg.
Camb.*, 1979, 82: 155–76, p. 159.

58. See A. Delaunay, *L'Institut Pasteur des origines à aujourd'hui*, Paris,
Editions France-Empire, 1962.

59. *ibid.*, see chapter 5.

60. Harriette Chick, Margaret Hume and Marjorie MacFarlane, *War on
disease: a history of the Lister Institute,* [London], A. Deutsch, [1971].

61. G. Wilson, *op. cit.* note 57, p. 157.

CHAPTER 11

1. 'Bonsi, Francesco. (1722–1803)', *Dizionario Biografico degli Italiani*,
vol. 12, Rome, 1970, pp. 382–3; G. W. Schrader, *Thierärztliches
Biographisch-literarisches Lexicon*, Stuttgart, Ebner and Seubert, 1863.

2. Francesco Bonsi, *Istruzione veterinaria pe'maniscalchi ecoloni sulla
presente epidemia contagiosa de'buoi*, Firenze, G. Piatti (*c.* 1800). For
Buniva's contemporary contributions, see chapter 5, pp. 83–5, and
notes 37–40.

3. Valentino Chiodi, *Storia della Veterinaria*, Milan, Farmitalia, 1957, pp.
446–7; E. Leclainche, *Histoire de la Médecine Vétérinaire*, Toulouse,
Office du Livre, 1936, p. 285.

4. Chap. 5, p. 70, note 8.

5. One who failed to fulfil his potential in this respect was Hunter's protegé William Moorcroft in London, *cf.* chapter 6, p. 95 and note 18.

6. Leclainche, *op. cit.* note 3, pp. 283–300.

7. Chapter 5, p. 83 and note 36.

8. *ibid.*, note 37.

9. Luigi S. Sacco (1769–1836) promoted Jenner's vaccination in Italy, using cowpox material from local sources in Lombardy.

10. Luigi Sacco, *Trattato di vaccinazione, con osservazioni sul giavardo e vajuolo pecorino*, Milan, Mussi, 1809.

11. Derrick Baxby, *Jenner's Smallpox Vaccine*, London, Heinemann Educational Books, 1981, p. 192; Sacco himself was doubtful and had been unable to reproduce the results; Sacco, *op. cit.* note 10, p. 178.

12. Genevieve Miller, *The adoption of inoculation for smallpox in England and France*, Philadelphia, University of Pennsylvania Press, 1957, p. 263.

13. *cf.* chapter 9, pp. 152–3 and notes 23–4.

14. Giovanni Pozzi (1769–1839) had completed doctorates in medicine and surgery, and served in the French army, when he became director of the veterinary school in Milan in 1807.

15. Leclainche, *op. cit.* note 6.

16. G. Pozzi, *Materia medica chimico-farmaceutica applicata all'uomo ed ai bruti*, Milan, Sonzogno e Compagni, 1816.

17. G. Pozzi, *Polizia degli spedali*, vol. 18 of: G. P. Frank, *Sistema compiuto di polizia medica*, Milan, Pirotta, 1830.

18. Agostino Bassi (1773–1856). A perspective on his life and work has been given by G. C. Ainsworth in *An introduction to the history of mycology*, Cambridge University Press, 1976, pp. 163–8.

19. A. Bassi, *Del mal del segno calcinaccio o moscardino*, Lodi, Orcesi, 1835.

20. See Ralph H. Major, 'Agostino Bassi and the parasitic theory of disease', *Bull. Hist. Med.*, 1944, 16: 97–107, pp. 102 and 103.

21. Leclainche, *op. cit.* note 3, pp. 301–5.

22. [—] Miessner, '150 Jahre der Hochschule Hannover', *Dt. tierärztliche Wschr.*, Sondernummer, June 1928, pp. 5–10.

23. *cf.* chapter 6, note 1.

24. *ibid.*, note 5.

25. Chapter 7, and *Med. Hist.*, 1977, 21, p. 20.

26. E. G. Lafosse, *Traité sur le véritable siège de la morve des chevaux, et les moyens d'y remédier*, Paris, Davis & Gonichon, 1749; Lafosse referred to the many valuable animals lost during 200 years of European wars; see also *Med. Hist.*, 1977, 1981, 25: 363–84.

27. Chapters 7 and 9.

28. L. Wilkinson, '"The other" John Hunter, M.D., F.R.S. (1754–1809): his contributions to the medical literature, and to the introduction of

animal experiments into infectious disease research', *Notes and Records R. Soc. Lond.*, 1982, **36**: 227–41.

29. K. H. Hertwig, 'Beiträge zur nähern Kenntniss der Wuthkrankheit oder Tollheit der Hunde', *Hufeland's Journal der practischen Arzneykunde und Wundarzneykunst*, 1828, **67**: 3–173. See also L. Wilkinson, 'Understanding the nature of rabies', in: *Rabies*, J. B. Campbell, K. M. Charlton (eds), Boston, Kluwer Acad. Publ., 1988, pp. 1–23.

30. Karl Heinrich Hertwig (1798–1881). Obit., *Arch. wissensch. prakt. Thierheilkde*, 1881, **7**: 495–8.

31. Ernst Friedrich Gurlt (1794–1882). Obit., *ibid.*, 1882, **8**: 486–501; see also G. Schuetzler, 'Das "Magazin für die gesammte Thierhielkunde" 1835–1874, die Herausgeber Gurlt und Hertwig und die Veröffentlichungen', *Berl. Münch. tierärztl. Wschr.* 1969, **82**: 81–6.

32. Brauell's work has been considered in more detail in chapter 7.

33. See Leon Z. Saunders, *Veterinary Pathology in Russia, 1860–1930*, Ithaca and London, Cornell University Press, 1980, pp. 19–35.

34. *Cf.* chapter 7, note 33. In 1862 Brauell published a volume on the pathology of rinderpest, see Saunders, *op. cit.*, p. 26.

35. Led to his earlier work on glanders by his interest in skin manifestations, Rayer never failed to consult veterinary colleagues when appropriate.

36. Thomas D. Brock, *Robert Koch*, Madison WI, Science Tech Publishers, 1988. Brock has perceptively analysed Koch's motivation throughout his career.

37. *ibid.*, p. 11.

38. E. H. Ackerknecht, *Rudolf Virchow: doctor, statesman, anthropologist*, Madison, University of Wisconsin Press, 1953, p. 108.

39. For Virchow's veterinary activities, see R. Völker-Carpin, 'Rudolf Virchow und die Veterinärmedizin', *Verhandl. xx. Int. Kongr. Gesch. Med. 1966*, Hildesheim, Georg Olms, 1968, pp. 558–95.

40. Chapter 8, p. 133 and note 9.

41. e.g., R. Virchow, 'Beiträge zur Kenntnis der Trichinosis und der Aktinomykosis bei den Schweinen', *Virch. Arch.*, 1884, **95**: 534–47, and **96**: 502–4.

42. [—], 'Professor Virchow on the deer disease', *Lancet* 1874, ii: 168–9.

43. Völker-Carpin, *op. cit.* note 39, p. 594.

44. R. Koch, 'Die Atiologie der Milzbrandkrankheit, begründet auf die Entwicklungsgeschichte des Bacillus Anthracis', *Beiträge zur Biologie der Pflanzen*, 1876, **2**: 277–310; *idem*, 'Die Aetiologie der Tuberculose', *Berl. klin. Wschr.*, 1882, **19**: 221–30.

45. R. Koch, 'Zur Untersuchungen von pathogenen Organismen', *Mittheilungen aus dem Kaiserlichen Gesundheitsamte*, 1881, **1**: 1–48.

46. Brock, *op. cit.* note 36, pp. 96–100.

47. See J. Gerlach, *Die Photographie als Hülfsmittel mikroskopischer Forschung*, Leipzig, W. Engelmann, 1863, pp. 2–5.

48. See Hyman Morrison, 'Carl Weigert', *Ann. med. Hist.*, 1924, **6**: 163–77; and Harold J. Conn, 'Development of histological staining', *Ciba Symp.*, 1945/46, **7**: 200–300.

49. L. Z. Saunders and C. N. Barron, 'A century of veterinary pathology at the A.F.I.P., 1870–1970', *Path. vet.*, 1970, 7: 193–224.

50. J. J. Woodward, 'The pathological anatomy and histology of the respiratory organs in the pleuropneumonia of cattle', *Report of the Commissioner of Agriculture on the diseases of cattle in the United States*, 1871, pp. 64–72.

51. Introducing the report to the Surgeon General, Woodward wrote: 'During the summer of 1869 the lungs of several cows, dead of epidemic pleuropneumonia, were brought to the Army Medical Museum by Professor John Gamgee, ...'. Gamgee's own paper was published in the same *Report* of 1871.

52. Brock, op. cit. note 36, chapter 22, 'An assessment of Koch and his work', and 'Introduction', p. 4.

53. *ibid.*, pp. 215–21.

54. As Paul Cranefield remarked, 'He was right then and he knew it. Later when he was wrong, he could not admit it, ...'. P. Cranefield, review of Brock's book, *Bull. Hist. Med.*, 1989, 63: 306–8.

55. Robert Koch, *Reise-Berichte über Rinderpest, Bubonenpest in Indien und Afrika, Tsetse-oder Surrakrankheit, Texasfieber, tropische Malaria, Schwarzwasserfieber*, Berlin, J. Springer, 1898.

56. Deborah Dwork, 'Koch and the Colonial Office: 1902–1904. The Second South Africa Expedition', *NTM-Schriftenr. Gesch. Naturw. Technik. Med.*, 1983, 20: 67–74.

57. R. Koch, 'Researches into the cause of cattle plague', *Br. med. J.*, 1897, i: 1245–6; also op. cit. note 55.

58. *ibid.*, p. 1246.

59. *Reise-Berichte*, op. cit. note 55, p. 1.

60. op. cit. note 58.

61. For a full account of the life and work of Arnold Theiler (whose son Max would one day receive a Nobel Prize for the development of a yellow fever vaccine), see the gushing, but exhaustive, biography by Thelma Gutsche, *There was a man. The life and times of Sir Arnold Theiler*, Cape Town, H. Timms, 1979.

62. Arnold Theiler, 'Die südafrikanische Pferdesterbe', *Dt. tierärztl. Wschr.*, 1901, 9: 201–3.

63. In London Stockman (1869–1926) worked closely with McFadyean; see I. Pattison, *John McFadyean*, London and New York, J. A. Allen, pp. 140–2; also Obit., *Vet. Rec.*, 1926: 498–9.

64. A. Theiler, 'Acute liver-atrophy and parenchymatous hepatitis in horses', *5th and 6th Rep. Dir. vet. Res., Dept. Agric. U.S. Afr.*, April 1918. In this paper Theiler also noted cases of apparent sexual transmission of serum hepatitis in the horse, p. 18–19.

65. Surra in cattle is caused by *Trypanosoma evansi* and is, like the trypanosomal disease of man, sleeping sickness, transmitted by blood-sucking flies, and possibly vampire bats.

66. *Reise-Berichte*, op. cit. note 55, p. 72.

67. Paul F. Cranefield, *op. cit.* note 54, p. 307; also P.F.C., 'East Coast

Fever: the second great cattle plague', paper given at symposium on 'history of animals and medicine' at Wellcome Institute, London, June 1989.

68. For later studies of the parasite of East Coast Fever, see E. V. Cowdry and A. W. Ham, 'Studies on East Coast fever. I. The life cycle of the parasite in ticks', *Parasitology*, 1932, **24**: 1–49; and G. Gettinby and W. Byrom, 'The dynamics of East Coast Fever: a modelling perspective for the integration of knowledge', *Parasitology Today*, 1989, **5**: 68–73.

69. D. Dwork, *op. cit.* note 56.

70. [F.] Loeffler and [P.] Frosch, 'Summarischer Bericht über die Ergebnisse der Untersuchungen der Kommission zur Erforschung der Maul- und Klauenseuche bei dem Institute für Infektionskrankheiten in Berlin', *Zentbl. Bakt. ParasitKde*, Abt. I, 1897, **22**: 257–9.

71. *ibid.*, p. 258.

72. E. Klein, 'The etiology of foot-and-mouth disease', *Lancet*, 1886, i: 15. It is tempting to see a connection here with Klein's simultaneous work, at the Brown Institution, on streptococci in milk causing scarlet fever, see L. G. Wilson, 'The historical riddle of milk-borne scarlet fever', *Bull. Hist. Med.*, 1986, **60**: 321–42.

73. [F.] Loeffler and [P.] Frosch, 'Berichte der Kommission zur Erforschung der Maul- und Klauenseuche bei dem Institut für Infektionskrankheiten in Berlin', *Zentbl. Bakt. ParasitKde*, Abt. I, 1898, **23**: 371–91.

74. W. D. Foster, *A History of Parasitology*, Edinburgh and London, E. & S. Livingstone Ltd., 1965, chapter 14, pp. 187–92.

75. *ibid.*, chapter 10, 'The trypanosomes', pp 115–35.

76. E. Perroncito, *La malattia dei minatori*, Turin, C. Pasta, 1910; R. Peduzzi, J. C. Piffaretti, 'Ancylostoma duodenale and the Saint Gothard Anaemia', *Br. med. J.*, 1983, ii: 1942–5; D. W. T. Crompton, 'Hookworm disease: current status and new directions', *Parasitology Today*, 1989, **5**: 1–2.

77. A. Delaunay, *L'Institut Pasteur des origines à aujourd'hui*, Paris, Editions France-Empire, 1962.

78. *Biographisches Lexikon hervorr. Arzte*, Berlin and Vienna, Urban & Schwarzenberg, 1932, p. 459.

CHAPTER 12

1. See 'Dr. Koch's newly described cholera organisms', *Br. med. J.*, 1883, ii: 828–9.

2. E. Lagrange, *Monsieur Roux*, Brussels, Goemaere, 1954, pp. 53–65.

3. e.g., James Lind, *Essay on diseases incidental to Europeans in hot climates*, London, T. Becket, 1768; J. Hunter, *Observations of the diseases of the army in Jamaica*, London, J. Johnson, 2nd ed., 1796; J. F. Lafosse, *Avis aux inhabitans des colonies, particulièrement à ceux de l'isle S. Domingue*, Paris, Royez, 1787.

4. W. D. Foster, A history of parasitology, Edinburgh and London, E. & S. Livingstone, 1965.

5. E. Roux et A. Yersin, 'Contribution à l'étude de la diphthérie', *Ann. Inst. Pasteur, Paris*, 1888, 2: 629–61; [E.] Behring and [S.] Kitasato, 'Ueber das Zustandekommen der Diphtherie-Immunität und der Tetanus-Immunität bei Thieren', *Dt. med. Wschr.*, 1890, 16: 1113–14; [E.] Behring and [E.] Wernicke. 'Ueber Immunisierung und Heilung von Versuchsthieren bei der Diphtherie', *Z. Hyg. InfektKrankh.*, 1892, 12: 10–44.

6. Albert Delaunay, *L'Institut Pasteur des origines à aujourd'hui*, Paris, Editions France-Empire, 1962, pp. 104–5.

7. Sir Rickman John Godlee, *Lord Lister*, Oxford, Clarendon Press, 1924, p. 496; the number of cases was originally mentioned in several of the letters below, notes 8 and 9.

8. The letters were printed, with the meeting's report, in *Pasteur Institute. Mansion House Fund*, London, J. Bale & Sons, 1889.

9. *Resolutions* from Mansion House Meeting, 1 July, 1889, on 'The Prevention of Hydrophobia'; Lister Institute Archives, catalogued by Lesley A. Hall, Contemporary Medical Archives Centre (CMAC), Wellcome Institute, SA/LIS, C.2. A number of letters from the original correspondence leading up to the meeting were given to the Lister archives in 1928 by the family of Sir James Whitehead, who was the Lord Mayor in 1889; SA/LIS C.1.

10. *Certificate of Incorporation of the British Institute of Preventive Medicine*, Lister archives, SA/LIS, C.7.

11. *Memorandum and Articles of Association, ibid.*

12. Harriette Chick, Margaret Hume and Marjorie MacFarlane, *War on Disease: a history of the Lister Institute,* [London], A. Deutsch, [1971].

13. See *The College of State Medicine* (In Liquidation), CMAC, SA/LIS D.7. and *The College of State Medicine*, Calendar for 1890–91, *ibid.*, p. 8; *College of State Medicine 1890/91, 1892, ibid.*, p. 9.

14. 'University of London. Meeting of the Senate', *Br. med. J.*, 1905, ii: 53.

15. Chick *et al.*, *op. cit.* note 12, pp. 135–6.

16. J. A. Arkwright, M. Burbury, S. P. Bedson, and H. B. Maitland, 'Observations on foot-and-mouth disease', *J. comp. Path. Ther.*, 1925, 38: 229–55; also, *ibid.*, 1927, 40: 5–30, and 79–117; F. C. Minett, *First Progress Report of the Foot-and-Mouth Disease Committee*, H.M. Stationery Office, 1925, p. 94.

17. H. M. Woodcock, 'Edward Alfred Minchin. 1866–1915', *Parasitology*, 1925, 17: 157–162; [—], 'A chair of protozoology in the University of London', *Br. med J.*, 1905, ii: 401; University of London Senate Minutes, ST/2/2/22, 101, 889, 1439–41, 1849–52, 2284–5.

18. e.g., E. A. Minchin and H. M. Woodcock, 'Observations on the trypanosome of the little owl (*Athene noctua*)', *Quart. J. Microscop. Sci.*, 1911, 57: 141–85; M. Robertson, 'Notes on the life-history of *Trypanosoma Gambiense...*', *Phil. Trans. R. Soc. Lond. B*, 1913, 203: 161–84.

19. The following year Manson became medical adviser to Joseph Chamberlain at the Colonial Office.

20. Chick *et al.*, *op. cit.* note 12, p. 94.

21. Lengthy reports were published by the Austrian Plague Commission in 1898, and by the German Plague Commission in 1899.

22. Discovered in Hong Kong by Yersin in 1894: Yersin, 'La peste bubonique à Hong Kong', *Annls Inst. Pasteur*, 1894, **8**: 662–7; also independently in the same year by S. Kitasato, 'The bacillus of bubonic plague', *Lancet*, 1894, **ii**: 428–30; see D. J. Bibel and T. H. Chen, 'Diagnosis of plague: an analysis of the Yersin–Kitasato controversy', *Bact. Rev.*, 1976, **40**: 633–52.

23. P.-L. Simond, 'La propagation de la peste', *Annls Inst Pasteur*, 1898, **12**: 625–87.

24. The constitution of the Advisory Committee, and of the working Commission in Bombay, is found in the introduction to the first report in the special Plague Numbers of the *Journal of Hygiene* published in September, 1906.

25. *ibid.* The reports, published over a period of 5 years, were accompanied by a number of illustrations showing the apparatus used, and also detailed anatomical drawings of the species of fleas involved in transmission.

26. See A. Delaunay, *L'Institut Pasteur des origines à aujourd'hui*, Paris, Editions France-Empire, 1962, pp. 104–5.

27. In June, 1898, the outbreak of bubonic plague which Yersin had studied in Hong Kong reached Nha Trang; see Yersin, 'Rapport sur la peste bubonique de Nhatrang (Annam)', *Annls Inst. Pasteur*, 1899, **13**: 251–61.

28. [—], 'Albert Calmette 1863–1933', *Annls Inst. Pasteur*, 1933, **51**: 559–64.

29. The existence of the toxins had first been recorded at the Paris institute in 1888, see *Annls Inst Pasteur*, *op. cit.* note 5.

30. See obit., *op. cit.* note 28, p. 560.

31. See E. Lagrange, *op. cit.* note 2, pp. 219–31.

32. Originally referred to as *Bacillus pestis*, the name was later changed to *Yersinia* after the discoverer.

33. Noel Bernard, *Yersin: pionnier-savant-explorateur (1863–1943)*, Paris, La Colombe, 1955.

34. For early examples, see G. W. Corner, *A History of The Rockefeller Institute 1901–1953*, New York, Rockefeller Institute Press, 1964, pp. 12–13.

35. The first privately endowed bacteriological laboratory in the United States had been a modest affair, founded in Brooklyn by a public-spirited physician in 1888, see René Dubos, 'Fess Avery: The Man and the Scientist', in: *Institute to University*, The Rockefeller University Press, 1977, p. 50. For Rockefeller contributions to British medical science and institutions see Donald Fisher, 'Rockefeller philanthropy and the British Empire: the creation of the London School of Hygiene and Tropical Medicine', *Hist. Educ.*, 1978, **7**: 129–43; and *idem*, 'The

Rockefeller Foundation and the development of scientific medicine in Great Britain', *Minerva*, 1978, 16: 20–41.

36. See Corner, *op. cit.* note 34.

37. From the 1908 Institute charter, quoted by Saul Benison, in 'Simon Flexner: the evolution of a career in medical science', in: *Institute to University, op. cit.* note 35, p. 21.

38. See Corner, *op. cit.* note 34.

39. Elizabeth Fee, *Disease and Discovery: a history of the Johns Hopkins School of Hygiene and Public Health 1916–1939*, Johns Hopkins, 1987.

40. Donald Fleming, *William H. Welch and the Rise of Modern Medicine*, Boston, Little, Brown & Co., 1954.

41. Paul Franklin Clark, 'Theobald Smith, student of disease (1859–1934)', *J. Hist. Med.*, 1959, 14: 490–514.

42. Theobald Smith and F. L. Kilborne, *Investigations into the nature, causation, and prevention of southern cattle fever*, 8th and 9th Annual Reports, 1891 & 1892, Bureau of Animal Industry, Washington, Government Printing Office, 1893; reprinted in: *Med. Classics*, 1937, 1: 372–597.

43. P. Manson-Bahr, 'The dawn of tropical medicine', *J. Trop. Med. Hyg.*, 1931, 34: 93–7; also, *idem.*, *Patrick Manson*, London, Thomas Nelson & Sons, Ltd., 1962, pp. 35–9.

44. *Med. Classics, op. cit.* note 42, pp. 471–526, and plate 10, p. 596.

45. P. Manson-Bahr and A. Alcock, *The life and work of Sir Patrick Manson*, London, Cassell and Co. Ltd., 1927, pp. 131–43.

46. W. Reed and J. Carroll, 'The etiology of yellow fever', *Am. Med.*, 1902, 3: 301–5; see also W. B. Dean, 'Walter Reed and the ordeal of yellow fever experiments', *Bull. Hist. Med.*, 1977, 51: 75–92.

47. Resuming the work with H. W. Graybill in 1920, Smith referred to the results of that first study undertaken for the Bureau of Animal Industry, H. W. Graybill and Theobald Smith, 'Production of fatal blackhead in turkeys by feeding embryonated eggs of *Heterakis papillosa*', *J. exp. Med.*, 1920, 31: 647–55, p. 647.

48. Theobald Smith, 'A comparative study of bovine tubercle bacilli and of human bacilli from sputum', *J. exp. Med.*, 1898, 3: 451–511, p. 510–11.

49. The Rockefeller Institute was certified by the State of New York in 1901, and the Board of Scientific Directors first convened in 1902; see Corner, *op. cit.* note 34, and Benison, op. cit. note 37, pp. 20–21.

50. Benison, *op. cit.*, who had access to Flexner's *Manuscript Autobiography* in the library of the American Philosophical Society.

51. *Op. cit.* note 46, and John M. Gibson, *Physician to the world.* University of Alabama Press, 1989.

52. Editor's introduction, *J. exp. Med.*, 1896, i: 1–3.

53. Simon Flexner and Eugene L. Opie, introductory page, *J. exp. Med.*, 1905, 7. See also Morris Fishbein, 'Some great medical editors', *Bull. Hist. Med.*, 1962, 36: 70–82, pp. 78–9.

54. Lloyd G. Stevenson, 'Bicentennial, Centennial and Semicentennial', *Bull. Hist. Med.*, 1976, 50: 1–3.

55. K. Landsteiner and E. Popper, 'Mikroskopische Präparate von einem menschlichen und zwei Affenrückenmarken', *Wien. klin. Wschr.*, 1980, **21**: 1830.

56. *idem*, 'Ubertragung der Poliomyelitis acuta auf Affen', *Z. Immun-Forsch. exp. Ther.* (Orig.), 1909, **2**: 377–90; S. Flexner and P. A. Lewis, 'The nature of the virus of epidemic poliomyelitis', *J. Am. med. Ass.*, 1909, **53**: 2095.

57. W. W. C. Topley, 'The spread of bacterial infection', *Lancet*, 1919, ii: 1–5; 45–9, 91–6; and *J. Hyg.*, 1920–1, **19**: 350–79; S. Flexner, 'Experimental epidemiology', *J. exp. Med.*, 1922, **36**: 9–14, and Harold L. Amoss, 'Experimental epidemiology. I. An artificially induced epidemic of mouse typhoid', *ibid.*: 25–69.

58. Paul Franklin Clark, 'Hideyo Noguchi, 1876–1928', *Bull. Hist. Med.*, 1959, **33**: 1–20.

59. Where he collaborated with its director in developing a protective serum against rattlesnake venom, see E. Schelde-Møller, *Thorvald Madsen*, Copenhagen, Nyt Nordisk Forlag, 1970, pp. 65–70.

60. N. P. Hudson, 'Adrian Stokes and yellow fever research; a tribute', *Trans. R. Soc. Trop. Med. Hyg.*, 1966, **60**: 170–4.

61. M. Theiler, 'Studies on the action of yellow fever virus in mice', *Ann. trop. Med. Parasit.*, 1930, **24**: 249–72.

62. Corner, *op. cit.* note 34, p. 87.

63. Sir Christopher Andrewes, 'Francis Peyton Rous 1879–1970', *Biogr. Mem. Fell. R. Soc. Lond.*, 1971, **17**: 643–62.

64. A. Borrel, 'Epithélioses infectieuses et épithéliomas', *Annls Inst. Pasteur*, 1903, **17**: 81–118.

65. V. Ellerman and O. Bang, 'Experimentelle Leukämie bei Hühnern', *Zentbl. Bakt. ParasitKde*, Abt. I, Orig., 1908, **46**: 595–609.

66. Peyton Rous, 'Transmission of a malignant new growth by means of a cellfree filtrate', *J. Am. med. Ass.*, 1911, **56**: 198. Rous was finally rewarded with a Nobel Prize in 1966.

67. Corner, *op. cit.* note 34, chapters 11, pp. 284–300, and 12, pp. 301–22.

68. David Bruce, 'Note on the discovery of a microorganism in Malta Fever', *The Practitioner*, 1887, **39**: 161–70.

69. Theobald Smith and M. Fabyan, 'Ueber die pathogene Wirkung des Bacillus abortus Bang', *Zentbl. Bakt. ParasitKde*, 1912, **51**: 549–55; T. Smith, 'Demonstration mikroskopischer Präparate von mit dem Bacillus des Abortus geimpften Meerschweinchen, *Berl. klin. Wschr.*, 1912, **49**: 715–16.

70. Theobald Smith, *Parasitism and Disease*, Princeton University Press, 1934.

71. Thomas Francis Jr., 'In honor of Richard E. Shope', pp. xxiii–xxxi, in: *Perspectives in Virology* IV, ed. M. Pollard, New York, Harper & Row, 1965; Obit., 'Richard Edwin Shope', *Lancet*, 1966, ii: 1033.

72. R. E. Shope, 'A filtrable virus causing a tumor-like condition in rabbits and its relationship to virus myxomatosum', *J. exp. Med.*, 1932, **56**: 803–22.

73. R. E. Shope, 'Influenza: history, epidemiology, and speculation', *Publ. Health Reports*, 1958, 73: 165–78.

74. *The Rockefeller Institute for Medical Research. History, Organization and Equipment*, New York, Rockefeller Institute, 1912, pp. 3–4.

75. Corner, *op. cit.* note 34, p. 133.

76. *ibid.*, p. 132.

77. In 1943 was published *Virus Diseases*, 'by members of the Rockefeller Institute for Medical Research': T. M. Rivers, W. M. Stanley, L. O. Kunkel, R. E. Shope, F. L. Horsfall, and Peyton Rous.

78. René Dubos, *op. cit.* note 35.

79. Frank Brink, Jr., 'Detlev Bronk and the development of the graduate education program'; and David Rockefeller, 'The University: climate of excellence', both in: *Institute to University, op. cit.* note 35.

80. Dubos, *op. cit.* note 35.

81. D. Fisher, *Minerva*, 1978, 16, *op. cit.* note 35, p. 20.

82. *idem, Hist. Educ.* 1978, 7, *op. cit.* note 35, p. 129.

83. N. Dungal, G. Gislason, and E. Taylor, 'Epizootic adenomatosis in the lungs of sheep – comparisons with jaagsiekte, verminous pneumonia and progressive pneumonia', *J. comp. Path. Ther.*, 1938, 51: 46–68; Björn Sigurdsson, 'Rida, a chronic encephalitis of sheep', *Br. vet. J.*, 1954, 110: 341–54, pp. 347–8.

84. B. Sigurdsson, 'Atypically slow infectious diseases', Extrait du Livre Jubiliare du Dr. Ludo van Bogaert, Brussels, *Acta Medica Belgica*, 1962.

85. See *J. exp. Med.*, 1943, 77: 315–22; and *ibid.*, 1943, 78: 17–26 and 341–5.

86. Pall A. Palsson, 'The Institute for Experimental Pathology at Keldur – The Early Years', in: *Björn Sigurdsson, dr. med. Collected Scientific Papers*, Reykjavik, Prentsmidjan Oddi, 1990, pp. xxii–xxviii, and personal communication. Sigurdsson's first paper from the 'Institute for Experimental Pathology' at Keldur was published in the *Journal of Bacteriology*, 1949, 58.

87. Björn Sigurdsson, 'Observations on three slow infections of sheep'. A series of Special University Lectures given in the University of London in March, 1954. Reprinted from *Br. vet. J.*, 1954, 110: 255–70; 307–22; 341–54.

88. *ibid.*, see pp. 44–5 (*Br. vet. J.*, 1954, 110, pp. 350–1).

88. G. R. Hartsough and Dieter Burger, 'Encephalopathy of mink, I and II', *J. infect. Dis.*, 1965, 115: 387–99.

90. D. Carleton Gajdusek, 'Unconventional viruses and the origin and disappearance of kuru', Nobel Lecture, 1976, in: *Les Prix Nobel en 1976*, Stockholm, Nordstedt & Sons, 1977, pp. 167–216.

91. *ibid.*, charts pp. 192–3; and P. A. Pálsson, 'Rida (scrapie) in Iceland and its epidemiology', in: *Slow transmissible diseases of the nervous system*, vol. I, eds S. B. Prusiner and W. J. Hadlow, New York, Academic Press, 1979, pp. 357–66.

Index

Abildgaard, P.C. 60, 69, 81
Académie de Médecine 147 ff.
Académie des Sciences 150 ff.
acarus (*see also* scabies) 139
'afflatus genitalis' 49
African horse sickness 133, 196
agents
    of infectious diseases 143, 144
    identification of 112
    multiplication of 137
agricultural societies 67, 78 ff.
agriculture 8
Albertus Magnus 18 ff.
Alembert, J. 67
Alexandrian library 7
Alfort (Paris) veterinary school 65 ff.
    teaching at 75, 90, 103
Amoss, H.L. 212
animalcula 48
animalcules 30–2, 85, 125
Andry, N. 48–9
animal experimentation 177 ff.
animal (veterinary) hospitals 5, 166 ff.
animal models 117 ff.
animal pathology 214, 217
Annales de l'Institut Pasteur 166
Annals of Agriculture 75
anthrax 7, 22, 123 ff.
    in blood of cattle and man 127
    in blood of sheep 126
    contaminated meat 124
    industrial 124
anthrax bacillus (rods) 126 ff, 192
    spores 129
anti-contagionism, anticontagionist
    117, 138
anti-vivisection, antivivisectionists
    166 ff.

Apsyrtus 13
Archives de Médecine comparée 111,
    123
Aristotle 6, 7, 13
Arkwright, J. 204
army surgeons 52
arthropod vectors 209
'Asiatic Cholera Medical Society' 124
Asklepios 5
attenuation 155, 173
Aurelius 8
Avery, O.T. 216
Avicenna 18, 22

Bacillus abortus Bang 214
bacterial viruses 179
    bacteriophage 179
bactéries see bacteridium
bactéridies see bacteridium
bacteridium 128–30, 155
bacteriology
    medical and veterinary 117, 130,
    173
    teaching 199
Baillie, M. 78
Baker, Sir George 93, 95
Balzac *père* (Bernard) 99
Bang, O. 213
Banks, Sir Joseph 59, 93
Barberet, D. 62
barber-surgeons 50
Bardsley, S.A. 82, 98
Barthélemy, Éloy 126, 150, 192
Bassi, A. 117, 188
Bates, Thomas 51–3
BCG vaccine 206
Bath and West of England Society
    88–9